# POPULAR SCIENCE BOOK OF HOME HEATING (AND COOLING)

**Evan Powell**

Southeast Editor
*Popular Science*

**Ernest V. Heyn**

Retired Editor-in-Chief
*Popular Science*

Reston Publishing Company
Reston, Virginia
*A Prentice-Hall Company*

## Important Notice About Prices and Sources

While this book was intended as a presentation of interesting products and methods and not as a catalogue, sources and approximate prices have been included whenever feasible.

All such items have been checked for availability and price at the time the authors delivered this book to the publisher. Some prices were not available, others are subject to change. Readers are requested *not* to write for information to the publisher, *Popular Science,* or the authors, but rather to the names and addresses in the captions or in the text.

Readers will note that some items do not include addresses in the captions or text; such items can be obtained at large, national retail outlets such as Sears, Montgomery Ward, Heath, Westinghouse, Carrier, etc.

The absence of prices is often caused by the variations in prices in different parts of the country, and sometimes the sale of exotic items only in limited areas.

**Library of Congress Cataloging in Publication Data**

Powell, Evan.
  Popular science book of home heating and cooling.

  1. Dwellings—Heating and ventilation.  2. Dwell-
ings—Air conditioning.  3. Dwellings—Energy con-
servation.  I. Heyn, Ernest Victor.
II. Title.
TH7224.P68  1983      697      82-18562
ISBN 0-8359-5564-8

Editorial/production supervision and interior design by
Ann L. Mohan

10   9   8   7   6   5   4   3   2   1

Printed in the United States of America

# Contents

# Foreword

**H**ome sweet home is especially sweet when it's warm in cold weather and cool in hot weather—a short summary of this book!

The table of contents is your best guide to problems you may have in either type of weather. In his introduction, my collaborator and highly qualified expert, Evan Powell, will tell you about his laboratory (actually the home he built from scratch atop a Carolina mountain), where many of the ideas and much of the machinery discussed in the book are tested under controlled conditions.

We have the privilege to adapt new and ingenious developments first reported in the pages of *Popular Science Magazine*. As former editor-in-chief, I have devoted my retirement years to projecting and producing, with staff experts like Evan, a series of books developed from the contents of the magazine. This is the sixth in the series.

The opening chapter contains information on the very latest in home heating—an exclusive look at the new pulse-combustion gas burner produced by Lennox and the newest catalytic wood stoves; also at programmed cooling—a microprocessor that controls in-window air conditioners. Next we talk about central heating—a choice between oil or gas furnaces or heat pumps. Then: thermostats, fireplaces, wood stoves, and a big

chapter on avoiding heat loss in your home (not neglecting the danger of home pollution, which is discussed in a chapter near the end of the book). Special attention is given to solar energy for heating the home, also to high-efficiency water heaters and to portable heaters as a supplement to your regular furnace. To add to our coverage of heat loss in the house, we include a chapter on weatherstripping. The final chapter is a useful checklist for home energy saving.

As you read the book, keep in mind that many chapters involve both heating and cooling problems, for instance: insulation, thermostats, weatherstripping, whole-house fans, heat pumps, and double glazing. Emphasis is on heating but includes cooling parenthetically, as we've done in our title and our first chapter heading.

As Evan will tell you, the book is based on "hands-on" experience—that is, we work with all the equipment we write about.

One final word before I turn you over to Evan: We couldn't have produced this guide to home heating and cooling without the support and cooperation of the people who run Times Mirror Magazines, of which *Popular Science* is an important part. My thanks, then, go to Jack Scott, president of Times Mirror Magazines; also to my successor as editor-in-chief of

*Pop Sci* (as it's affectionately called), Hubert P. Luckett ("Lucky"), who is now vice-president and editorial director of the company; Ken Gilmore, current editor-in-chief of *Pop Sci*; and the many members of its staff, whose advice and expertise I frequently called upon, among them: two Group Editors, Al Lees and Rich Stepler; Art Editor Dick Meyer; etc., etc. Also an appreciative nod to Marita Begley and Eleanor Freebern, who contributed many hours of research. A nod of enthusiastic approval, too, for the splendid editorial handling of the book by Ann Mohan at Reston Publishing Company.

I acknowledge with profound thanks the cooperation of Jim Liston, editor of *Homeowners How-To*, another Times Mirror Magazine which has published many pages of informative material by Evan Powell. While specific credit to Jim's magazine has been unwieldy, I want to add this record of my appreciation of the splendid text and art he permitted Evan and me to include in this book.

Our publishers also acknowledge the cooperation of all these people and regret that there is only room for two bylines, mine and that of my good friend and collaborator, Evan Powell

*Ernest V. Heyn*

# Introduction

Bernice Roberts was the teacher of my seventh grade class at Swannanoa High School in North Carolina and was one of those rare persons in that field who managed to keep learning interesting, exciting, and fun. That special knack—and the uncanny way her admonitions to us came true—are why most of the points she tried to get across to us really stuck. She always looked ahead; in those days, her big thing was audiovisual equipment. Our class was the "camera crew" for the entire school and the envy even of some high school seniors. She made us feel important. If she were transported into today's teaching world, you can bet that all her students would have at the very minimum a 64K computer terminal at their desks. Somehow she would get them!

She looked to the future in other ways, too. The admonition I remember with greatest clarity was one that she emphasized to us over and over. "Within your lifetime," she would say, "you'll pay a dollar a gallon for gasoline and fuel oil, and it will be hard to get. There's only so much, and it can't last forever." Bernice Roberts' insight qualified her not only as a teacher but as a prophet, because it was in 1950 that she was sounding this warning.

Surely this elementary school teacher in the mountains of North Carolina was not the only brilliant mind in this country who recognized the coming energy crunch. Why, then, the shock of 1974? Many reasons can be cited: complacency, indifference, lack of foresight. Twenty-five years after Bernice Roberts' prophetic statement, our homes were literally caught with their doors open. Cold air entered readily through cracks and crevices, and heat flowed out uninsulated walls and ceilings. There had never before been a real incentive to build those homes to conserve that energy. Insulation, good weatherstripping practices, double-glazed windows; all these cost far more than would be saved in many years through reduced energy costs. Think about this: Most of us today still live in the houses of that period.

The ever-increasing costs of energy have served to keep a very real incentive before us. We've insulated and caulked and weatherstripped. We've experimented with various types of fuels, and we now have available to us a number of good alternatives for high-efficiency heating and cooling. The energy crisis has also created an incentive for "snake oil" salesmen and con artists to step in

Evan Powell's *Popular Science* experimental house, Chestnut Mountain, on a mountain top above Travelers Rest, South Carolina.

with anything they can devise in the name of such conservation; whether it works or not is inmaterial to them as long as it is profitable.

To the homeowner who picks and chooses wisely, there are great rewards in store. Some have won the battle of the energy crunch and are actually heating and cooling their homes for less than they were in 1972. But every home and every lifestyle is different. The energy needs of those homes are also different, as are the combinations of equipment, techniques, and routines necessary to meet those energy needs while still maintaining comfort. The intention of this book is to

help you separate the wheat from the chaff. There are a number of excellent devices on the market today that can help you realize and possibly exceed the energy goal you may have set for your own home; there are others that may be a complete waste of time and money. This book is no mere review of gadgetry where we read product specifications and say "This should work okay," or "This won't work." This is equipment and these are techniques that we have used and lived with and found to work for us.

It has become apparent to me in evaluating all types of home equipment that the important aspect is not to "get numbers" that tell you of the efficiency or power consumption or heat output of the particular system or piece of equipment. Most heating and cooling appliances are tested in laboratories to specific standards, and that information is readily available. Rather than duplicate that work, we ask, "How will it actually work in a home? What are some of the advantages and disadvantages that may not show up in lab tests?" If one type of wood stove, for example, rates a few percentage points higher than another in an efficiency test but is inconvenient (or dangerous) to use then it may not be the best choice. What about the salesman who tells you that the heat from a wood stove will distribute itself evenly throughout your house no matter where you put it? Is the heat from a wood stove so different from that produced by a furnace, which must rely on an elaborate distribution system of ducts, dampers, and grilles to distribute the heat evenly? And if the heat from your wood stove or solar collector doesn't distribute itself evenly, what can you do about it?

In order to provide a workshop where all types of home equipment can be examined to provide true and practical answers to readers, consumers, and manufacturers, the Chestnut Mountain project was conceived. Situated on a mountaintop above Traveler's Rest, South Carolina, Chestnut Mountain (see photo) is a real home—I know because I live there. But it

does a lot more than provide living and working quarters. From the ground up, a primary part of the concept was that every portion of it, including the structure itself, must be practical and completely replicable by any reader. As you'll see, we've never veered from that basic concept except in some instances where we're working with equipment six months to a year before it is available publicly so we can compile information and research by the time the product is marketed. While laboratory and research testing is not a goal, some amount of monitoring is necessary, so the home is completely instrumented for that purpose. A Heath weather computer tabulates data on weather conditions for comparing heating and cooling equipment performance to degree days.

For the most part, the products and projects that you'll find in this book are those that we have lived with in the Chestnut Mountain project. We have installed and monitored them under real-life situations and conditions and found that they did, indeed, work. With appropriate adjustments to fit your own home and lifestyle, they should also work for you in reducing the energy consumption of your home for heating and cooling.

Bernice Roberts would have loved the challenge of the world of the 1980's and the Chestnut Mountain project.

*Evan Powell*

# WHAT'S HOT IN HOME HEATING (AND COOLING)

Only two years ago this chapter would have been impossible—or at least wouldn't contain much information that was really "hot." Why? Despite our emphasis on developing new equipment for more efficient energy utilization, the equipment—"hardware" in today's terminology—still wasn't available. The "energy-efficient" equipment that we had to heat and cool our homes was for the most part highly-tuned modifications of the same types of equipment we used twenty years ago.

Editors at *Popular Science* are very fortunate in that they often get an advance look into the hardware of the future, and what they see is exciting for the homeowner who is seeking to reduce the cost of operating his home—and that includes all of us.

There are three major factors involved in an all-inclusive program for reducing your home's operating costs: the heating/cooling equipment, the control systems, and the techniques. We have selected the "hottest" examples of each of these categories for this chapter. As you read, think in terms of the individual parts that can be applied to your home.

If your home's heating system will soon be in need of replacement, the newest example of high technology is a home furnace that engineers have dreamed of for decades, a "heat engine" that is so efficient that the heat left over from the fuel is so minimal that it can be exhausted through the wall in a plastic pipe. Meet the Pulse-Combustion Furnace!

# A 96% efficient pulse-combustion furnace

*With 70 tiny explosions per second, this new gas furnace captures nearly all the fuel's heat. A bonus: It needs no chimney*

## By Evan Powell

Even though the countdown was going smoothly, the drama was intense. As the count reached T minus 30 seconds there was a slight whirring noise. Then, right on schedule, we had ignition.

This blastoff didn't occur at Cape Canaveral but in the basement of a house in Greenville, South Carolina. I wasn't in the company of aerospace engineers, but with the city heating inspector and officials of the Piedmont Natural Gas Co. The machine was not the space shuttle, but one that may prove to be of more immediate significance to homeowners—a new type of heating system, the Lennox pulse-combustion furnace.

Inside the modest beige cabinet before us was a burner that works more like your car's engine than like a conventional furnace burner. Very small charges of gas

and air are admitted to the combustion chamber where they ignite and burn, producing 60 to 70 mini-explosions every second (see box, page 4). Result: Combustion is so complete that nearly all the fuel's heat energy is released.

Furthermore, the furnace's heat-exchanger network extracts most of the heat from the exhaust gases. In a conventional furnace, these gases must remain above 300 degrees F (they're usually much hotter) so that the chimney will draw properly and water vapor in the gases won't condense. That can represent a great deal of lost heat. But the Lennox Pulse furnace has no chimney—flue gases are vented through a small plastic pipe. The force generated by the combustion process pushes them out. The flue gases are cooled to less than 100 degrees, the water vapor condenses, and its latent heat (released when a gas changes to a liquid) is captured. The heat exchanger is specially designed to handle the condensate, which then flows away through a drain line.

The pulse-combustion concept is not new. In fact, a boiler, the Hydropulse, made by Hydrotherm, Inc., in Northvale, New Jersey, has been on the U.S. market for about three years. But a pulse-combustion furnace is news. (A furnace heats by circulating warm air, a boiler by circulating hot water or steam.) The Pulse furnace I was installing was the first one Lennox had released for home testing without company supervision.

## INFAMOUS ANCESTOR

Making a practical pulse-combustion furnace is trickier than making a boiler. The reason: The basic combustion process is noisy—it's derived from the technology that gave us the German V-1 buzz bombs of World War II. Without the damping effect of the boiler's surrounding water jacket, that sound is difficult to soften.

In the early '60s, the American Gas Association (AGA) Laboratories in Cleveland, under a Gas Research Institute contract, began development work on a pulse-combustion furnace. Eventually, AGA Labs sought a furnace manufacturer to assume the project. Lennox was the only taker. In 1976, AGA gave Lennox two crude working models of the Pulse furnace. "There were two good things about them," says Tom Morton, one of the chief project engineers. "They worked, and they were efficient. But the people in the lab had to wear earmuffs to tolerate the noise."

To bring the noise to an acceptable level, the engineers experimented with different materials and different weights of materials for the critical components. Each time one was changed, they had to back up and see how that affected all the others. Slowly they arrived at a satisfactory combination. They also tinkered with the placement of a baffle in the tailpipe and a resonator to reduce sound levels at certain frequencies. Finally, they mounted the combustion-chamber and heat-exchanger assemblies on damping pads and insulated the cabinet with fiberglass.

In 1980, some prototypes were installed in homes for field tests—with very

## How It Works

When the burner is energized, gas and air in precisely metered amounts enter the sealed combustion chamber (1) through separate ports (2) and (3), each backed by a flapper valve. A spark-ignition device (4), much like the spark plug in your car, fires the initial charge, creating the first pulse (5). Each pulse burns only about 0.003 cu. ft. of gas, equivalent to ¼ to ½ Btu. The pressure from the pulse closes the flapper valves, forcing combustion products to travel through the tailpipe (6), the only remaining exit from the chamber. The pulse travels on past the resonator (7), a noise-control device, and into the exhaust decoupler (8), which contains a stagnant air mass that acts as a baffle or cushion and reflects part of the pulse back through the tailpipe into the combustion chamber. In the meantime, a partial vacuum in the chamber, left by the displaced gases, has allowed the flapper valves to open again and fill the chamber with a second charge of fuel and air. The returning pulse ignites the mixture and the cycle repeats. (The igniter plug is used only for the first pulse.)

The force of each pulse creates great turbulence, causing maximum heat transfer and pushing the combustion products through the entire heat-exchanger network. Within the exhaust decoupler, the gases are cooled from 1,200 deg. F to about 350. They are then forced through a fin-and-tube coil (9), where they are cooled to less than 100 deg. As the temperature reaches the dew point (around 130 deg.) inside the coil, water vapor condenses and the furnace captures the latent heat of the phase change. The remaining flue gases exit through a PVC tube, and the condensate drains away (10). Throughout this time, supply air from the house passes across all these components (11) designed for maximum heat transfer, absorbing most of the heat.

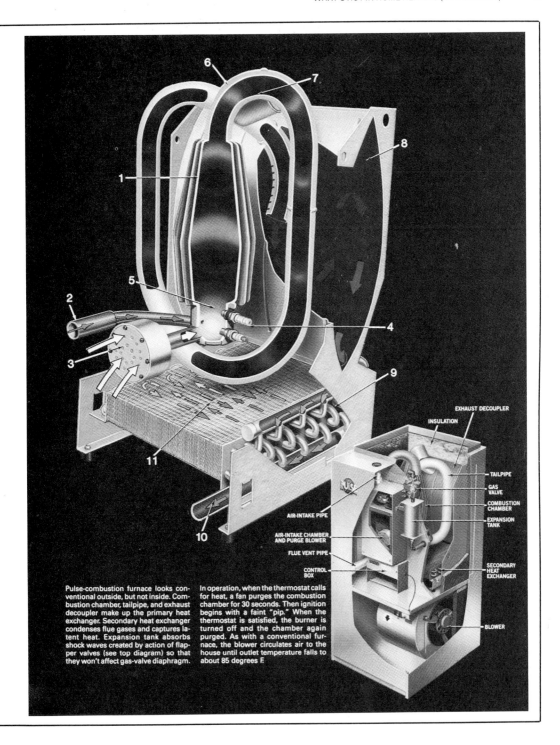

EXHAUST DECOUPLER

INSULATION

TAILPIPE

GAS VALVE

COMBUSTION CHAMBER

EXPANSION TANK

AIR-INTAKE PIPE

SECONDARY HEAT EXCHANGER

AIR-INTAKE CHAMBER AND PURGE BLOWER

FLUE VENT PIPE

CONTROL BOX

BLOWER

Pulse-combustion furnace looks conventional outside, but not inside. Combustion chamber, tailpipe, and exhaust decoupler make up the primary heat exchanger. Secondary heat exchanger condenses flue gases and captures latent heat. Expansion tank absorbs shock waves created by action of flapper valves (see top diagram) so that they won't affect gas-valve diaphragm. In operation, when the thermostat calls for heat, a fan purges the combustion chamber for 30 seconds. Then ignition begins with a faint "pip." When the thermostat is satisfied, the burner is turned off and the chamber again purged. As with a conventional furnace, the blower circulates air to the house until outlet temperature falls to about 85 degrees F.

good results. In fact, at a house in Cleveland where one was installed that fall, the gas-meter reader knocked on the door in January and asked if the family had been away for the winter. Seems their gas bill was less than half what it normally should have been. When he learned the reason, the meter man called Lennox to see if he could get a pulse-combustion furnace for his own house.

Tests bear out such anecdotal evidence of the furnace's efficiency. Fossil-fuel heating systems are rated by Btu-per-hour input (the heat available in the quantity of fuel burned in an hour) and output (the heat actually extracted to heat your house). Typically, a high-efficiency gas furnace of 100,000-Btu/h-input rating will have an output of about 70,000 Btu/h. The difference between input and output goes up the chimney in the form of heat, unburned fuel, and water vapor.

Furnace technicians accept such inefficiency and size heating systems accordingly. But the pulse-combustion furnace changes the rules, as I realized when conferring with David Chase, Lennox's director for government and public relations, about the size of my test furnace. He suggested an 80,000-Btu/h-input model. I immediately calculated about a 56,000-Btu/h output. "Sounds too small," I replied. "What's the output?" "Just about 80,000 Btu," Chase replied with a grin.

The Pulse furnace's annual fuel-utilization efficiency (AFUE), determined according to Department of Energy standards, is 91 to 96 percent, depending on the model. As the name implies, this rating is designed to measure efficiency on an annual (or seasonal) basis. It takes into account not only the percentage of the heat energy captured in the combustion process, but also off-cycle heat loss: the heat that wafts up the chimney while the burner is off. That can be a good bit, since most furnaces draw combustion air from

the house and continue to vent heated air up the chimney when the burner is off. The Pulse furnace has a closed combustion system: Combustion air is piped in directly from outside, and, of course, there is no chimney. Thus its off-cycle losses are minimal.

If gas furnaces now in American homes were rated according to the AFUE tests, says Lennox, 30 million would be shown to operate at only 50 to 60 percent efficiency. The Pulse furnace is also notably more efficient than new conventional furnaces: Lennox's own most efficient model (with vent damper and direct-spark ignition) has an AFUE of 78.6 percent—and it's one of the most efficient available. New furnaces at the bottom of the list on a Department of Energy fact sheet rate a measly 55 percent AFUE.

Put in terms of dollars, if your annual fuel bill is $1,000 and your furnace operates at about 66 percent efficiency, the Pulse furnace would reduce your bill by more than $300. If your furnace is only 55 percent efficient, you'd save around $425. Even if you have one of the most efficient conventional furnaces, you could save about $180 per season.

Lennox expects the Pulse furnaces to sell for around $1,800 installed (though that depends on the ductwork required). A conventional high-efficiency furnace might have an installed cost of $1,200 to $1,600. With the energy savings, however, the Pulse furnace would repay its extra cost within two years, says Lennox.

And since it needs no chimney, the Pulse could actually be less expensive to install in a new house than a conventional fossil-fuel furnace—the chimney can be one of the major costs of a heating system. If your chimney needs repair, the Pulse furnace could also prove a less-expensive alternative. And it allows you to locate the furnace to minimize duct runs or to put it in an out-of-the-way place.

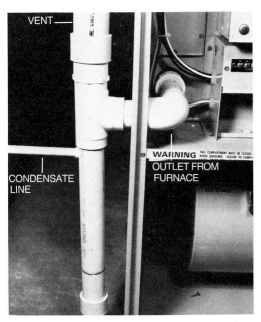

**Low-temperature exhaust residue** requires no chimney, is carried to outside of building through 1½-in. plastic pipe. Small, ½-in. pipe attached at this point drains condensate water away.

**Gas and air enter** combustion chamber through separate flapper valves. Orifice at gas flapper valve, shown here, determines amount of gas that can enter at each pulse.

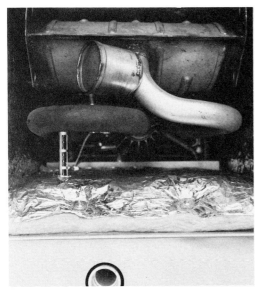

**View from the top.** Configuration of airfoil-shaped decoupler is evident at top of photo. "Can" attached to tailpipe is resonator that absorbs certain sound frequencies and reduces noise level.

**Pulse burner** at right is dwarfed by oil-fired furnace which previously supplied heat for test home. At latest firing rates, output of each is the same.

## EASY TO LIVE WITH

The Lennox Pulse can replace many newer furnaces—without modification to plenum or duct systems—as well as almost all warm-air systems with a little modification.

Such accessories as humidifiers and air cleaners work with the Pulse furnace, and its direct-drive blower will handle air-conditioning coils in nominal ratings of up to five tons.

The thermostat is conventional, and you can use a setback thermostat to reduce your heating costs even more. Internal controls are much the same as for ordinary furnaces, too, so service problems should be minimal.

Installing the Pulse furnace is similar to putting in a conventional furnace except for the venting and condensate drain. A 1½-inch polyvinyl chloride (PVC) pipe connects to a drain-leg fitting. From the top of this fitting a 2-inch PVC pipe takes the flue products vertically or through a wall to the outdoors, while from the bottom a trickle of condensate flows through a small PVC tube to a floor drain. The condensate is slightly acidic but noncorrosive to plastic, tile, cast sewer pipes, and septic tanks. Outdoor air for combustion is piped through another PVC pipe.

Knowing its buzz-bomb ancestry, I was especially curious about how much noise the furnace made. It turned out to be far quieter than I expected, though louder than a simple atmospheric gas burner. The sound is a sort of hum, similar to that of a powered gas burner or an oil burner but with a muffled "popping" instead of the rumbling.

The sound frequency of the Pulse furnace is such that certain duct systems could pick up harmonics and amplify them. For that reason Lennox insists on sound-isolation techniques that include using pads (furnished) under the cabinet, flexible canvas boots on the plenums, a flexible gas line connector (furnished), and foam insulation on the supports for the plastic vent and intake tubing. Thus installed, the Pulse furnace could be quieter than the one it replaces.

Lennox claims that the Pulse is also the safest furnace ever produced. Safety controls shut it down long before combustion products could escape. A flame-rectification sensor verifies ignition and locks out the gas valve in case of failure. The burner will also shut down immediately if the air intake or exhaust outlet is obstructed.

The Pulse furnace is AGA-approved and Underwriter's Laboratories listed, but if you're the first in your locality to install one, you may have to get a zoning variance to vent it through a PVC pipe.

For more information you can contact Lennox Industries Inc. (Dept. PC, Box 400450, Dallas, Texas 75240). The Pulse furnace is available through Lennox dealers nationally.

Pulse combustion is one high-technology answer for the homeowner who needs to replace his present heating system. But what about those of us who want to upgrade what we have? You'll find more answers to that in Chapter 2, "Central Heating—Oil, Gas, or Heat Pumps?"

Another route that many homeowners have chosen is to use the approach of alternative-fuel heating, primarily wood heating. In the past, that course has been fraught with obstacles. Conventional wood stoves aren't very efficient, contain a considerable amount of pollutants in the exhaust, and can create accumulations of creosote that can lead to chimney fires. Nothing new about these problems—they have been around from the beginning with "airtight" wood stoves. But suddenly, modern technology steps in with an answer in the form of a catalytic device that burns the pollutants in wood smoke at a lower temperature than was ever before possible. In chemical terms, rather than HC (hydrocarbons) and CO (carbon monoxide, a toxic gas) coming out the chimney, there is $CO_2$ (carbon dioxide, abundantly present in nature) and $H_2O$ (water vapor). Seem too good to be true? It does, but happily it is true, although too few are aware that it exists. Here's what you should know to select and use a catalytic wood stove:

# How to get every last Btu (safely) from your new catalytic wood burner

---

*Proper use isn't difficult, but it* is *slightly different*

---

## By Evan Powell

Once upon a time in the land of America, the fuel that the people used to heat their homes became scarce and very expensive. In many areas they turned to stoves that burned the wood from the forests around them. But soon they learned that the stoves caused a tarry accumulation that could start dangerous fires in their chimneys. So the craftsmen drilled some holes in the stoves to change their air flow. "No more creosote," the craftsmen cried, and the people bought more stoves.

But chimney fires continued, and it also became apparent that the wood stoves were not very efficient and used more wood than they should. The craftsmen made more alterations. "No creosote and

high efficiency," they cried, and again the people came and bought the stoves. But chimney fires and low efficiency persisted, and there came another dire warning: The stoves were breathing poisonous gases into the air.

As the craftsmen were wringing their hands, a white knight rode up with a magic amulet for the stoves. "Build your wood stove around this and you'll protect yourself from the evils of creosote and poisonous fumes—and get far better efficiency to boot," he told the craftsmen. So a number of them took the amulet, built their wood stoves around it, and then cried, "No more creosote, no more toxic fumes, and the highest efficiency yet." But by now the people had wearied of the craftsmen's cries and few of them came to buy the wood stoves with the magic amulet.

With apologies to Aesop, this modern version of "The Boy Who Cried Wolf" is no fable. The amulet is the catalytic combustor.

During the past two years I've worked with more than a dozen models of cata-

**Catalytic wood stoves** require far less fuel than conventional counterparts, but you'll get best results from regular routine of "feeding." Webster Oak model shown retains ornate 19th century look, complete with foot warmer rails, but adds catalytic efficiency.

lytic stoves, including the latest models with improved features, and I'm very enthusiastic about them. But life with a catalytic wood stove is quite different from what you may be used to. Before you buy one, you should be aware of their special use procedures and limitations.

## FIRST ENCOUNTERS

I suppose I looked a little funny dressed in a coat and tie clambering up a stepladder beside a big tent, erected on the roof of the Hyatt Regency Hotel in New Orleans a couple of years ago. My goal: to poke nose and hands into three flue pipes projecting from the tent's roof. The occasion was the international trade show of the Wood Heating Alliance, an organization of makers of wood-burning appliances. Inside the tent were three wood stoves that had been burning—slowly and continuously—

for three days. And getting a first-hand look into their flues seemed more important than decorum—more, even, than my usual fear of heights.

What I found was worth the discomfort. The flue liners were smooth, with hardly a smudge of soot. There was none of the tar or creosote you'd expect after a prolonged low-burn period. The reason: The three stoves contained an intriguing new element—a precious-metal-coated ceramic honeycomb disc about six inches in diameter and three inches thick. The disc is a catalytic combustor, developed especially for wood-burning stoves by Corning Glass Works, which also developed the catalytic converter for your automobile. Used in a wood stove, the catalytic combustor is a triple-threat device that not only reduces creosote, as the flues atop the tent attested; it also increases the efficiency of the stove and reduces air pollution.

As I write, there are more than 20 manufacturers making wood stoves,

**Secret of Webster's high efficiency** is excellent design of this 100-pound combustor housing which serves as heat sink. Finned inside and out, it holds heat for hours after fire subsides. Surface thermometer monitors combustor performance.

**Catalytic wood stoves** look conventional, but most include a viewing port for monitoring the combustor. Lavec's port is in the teakettle's shadow (top). Catalytic combustors (bottom) make flammables in wood smoke burn at a lower-than-normal temperature. The cells at left in photo are ¼-in. squares; the experimental disc at right is ⅛-in. honeycomb.

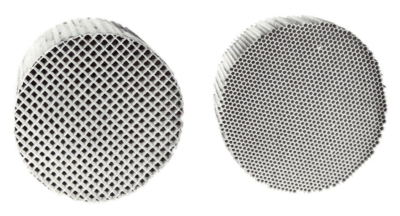

wood-burning furnaces, and fireplace inserts using Corning's catalytic combustor, a similar one made by Applied Ceramics, or one of foreign origin. Solar Key uses a precious-metal-coated catalytic steel plate, and other companies may soon get in the act. Penn Stoves makes an add-on heat exchanger using Corning's combustor. It can be put in the flue pipes of top-vented wood stoves.

## HOW THE COMBUSTOR WORKS

Under normal conditions, hydrocarbons and other potentially flammable products in wood smoke have a wide range of ignition temperatures. Many won't burn below about 1,200° F or even higher. In conventional stoves, the average temperature in the firebox is only 700 to 800°. Thus many of those products pass through unburned, wasting a good bit of the potential heat in the wood. Worse, some of the unburned products condense on the flue interior as creosote, the fuel that fires chimney blazes. Others waft out the chimney and pollute the air.

In a catalytic stove, the combustor disc is placed in a special chamber at the top of the firebox (see diagram) and a second fire burns there, using the byproducts from the wood fire below as fuel. A catalyst is a substance that can trigger a chemical reaction and allow it to take place under milder conditions than normal; the catalyst itself is unchanged at the end of the reaction. The catalyst used on the combustor is a thin metal coating (usually platinum, palladium, or a combination) that allows nearly all the hydrocarbons and other flammable products in the smoke to burn at a temperature much lower than usual.

"There is still disagreement even among scientists as to exactly what happens," says Roger Allaire, senior project engineer for Corning. But one widely held theory is that the catalyst sets up a tug of war, on the atomic level, with substances in the wood smoke. "The atomic bonds of crystalline materials, such as platinum and palladium, are unbalanced at the surface," Allaire explains. "This tends to strain the bonds of other substances that come near—in this case, stuff in the wood smoke. With this strain on the bonds, they will break more easily, meaning the substances will burn more readily—at about 450 degrees F, which is easily attained soon after the fire is lit."

Once the catalytic reaction begins, the hydrocarbons start burning—mostly within the combustor. This causes the temperature there to soar, sometimes to about 2,000°, which makes it work even better. Ultimately most of the flammable substances burn, and much of the wood's Btu content that is normally wasted is captured: The residue that otherwise would create creosote and air pollution creates heat instead.

Tests conducted by independent laboratories indicate that the catalytic combustor reduces creosote formation by up to 90 percent, compared with conventional airtight wood stoves, and substantially reduces air pollution, a less-publicized problem of the wood-burning renaissance.

More than seven million wood stoves have been sold since 1973, according to the Environmental Protection Agency, and home stoves now provide the heat equivalent of 140,000 barrels of oil a day. But at a price: "The rapidly increasing use of wood stoves is causing concern about the effect on air quality," says John Milliken of the EPA. The smoke pall can be very heavy over mountain valleys in the Rockies and New England. With a catalytic stove, the hydrocarbons and carbon monoxide in the smoke are mostly oxidized to become carbon dioxide and water vapor, and particulate emissions are reduced by up to 66 percent, according to tests.

**Timberline T-Cat-I** uses new rectangular combustor, has viewing window located in front of combustor. U-shaped heat exchanger above combustor aids heat transfer. Blower is optional.

**Birmingham Ponderosa Catalytic** has viewing window located above combustor on outlet housing. Firebox bottom is insulated for early combustor lightoff. Bypass damper operates auotmatically when door is opened.

**Riteway catalytic** is designed for function, and it succeeds. Bimetal thermostatic damper controls firing rate. Large heat exchange area gave consistent heat output in our basement test installation. Shaker grates quickly clear ashes from combustion chamber.

**Thermostatic combustion air control** on Royale Scot is adjusted as needed to maintain skin temperature of stove. Author found that any surface temperature above 250 degrees F maintained combustor operation properly, resulted in consistent room temperatures. Air passes under firebox to preheat it before entering chamber.

Efficiency improvements are also impressive. Auburn University's Wood Heating Laboratory rates a very good non-catalytic airtight stove at about 55 percent heating efficiency, comparing the heat output to the room with the Btu content of the wood. Certicon of America, Inc., rated the Franklin Concorde (no longer in production, but essentially similar to the current Lavec 2000) at 83 percent heating efficiency. The Concorde's combustion efficiency ran about 94 percent. Combustion efficiency measures the heat released within the stove in Btu as a percentage of the Btu content of the wood. It does not consider the stove's ability to send the heat out into the room rather than up the flue.

At Canada's Ryerson Polytechnical Institute, the Lakewood Unicorn Catalytic stove showed a combustion efficiency of 94 percent and a heating efficiency of 75 percent. The same stove, when tested without the catalytic combustor, showed a combustion efficiency of 86 percent and a heating efficiency of 64.

"It's difficult to compare results between the six or seven laboratories presently testing wood-burning stoves since each uses a slightly different test method," says Lakewood's president, Clyde L. Logue. "But most of the labs are finding an increase of 15 to 25 percentage points in heating efficiency using a catalytic combustor." That's significant.

## IN-HOME RESULTS

I was even more impressed after my own three-month initial test of the catalytics. A "white glove" test of the interior of the flue revealed no measurable creosote buildup. I should also mention that buildup was only about 3/32 inch thick after two season's use of my non-catalytic Buck stove. I attribute that to my habit of opening the dampers and allowing the stove to

burn hot for about 15 minutes each day. With a catalytic stove that chore is not only unnecessary, it's useless, since the stack wouldn't get hot enough to result in any cleaning action even if there were something to burn off.

Outdoors, the smell of wood smoke was diminished substantially with the catalytic stove, indicating reduced pollution. I had been warned that I might smell sulfur, as with some automotive catalytic converters, but I never have.

## DURABILITY

Nobody really knows how long catalytic combustors will last. Most experts say at

**Steel King central boiler** can be used as supplemental boiler, is also available in warm-air furnace version. Upper portion of firebox is designed especially for catalytic combustion.

least three years; some say ten years or more. Most are warranted for a year.

When should you change one? Corning's Susan Kalian advises: "If the combustor will fire to a high-visibility temperature and hold it for several minutes, it's okay." A replacement disc may cost $60 to $80, but the price could drop as production increases. On most catalytic stoves, changing the combustor is a simple matter of removing a couple of stainless-steel pins and allowing it to drop free. With some stoves it can be removed for coal burning. Solar Key will replate the catalytic combustor in its furnace at moderate cost.

I found that the high heat generated within the combustor makes it effectively self-cleaning. Though I tried, I could not

build up a permanent accumulation of gunk in the ¼-inch cells of the Corning combustor, or even in Applied Ceramics' unit with ⅛-inch cells. "Our tests, and those of independent laboratories, indicate that under normal conditions in a well-designed stove, cell plugging won't likely occur," says Kalian.

## CATALYTIC CARE AND FEEDING

Remember the basic principles of catalytic combustion and you'll have no problem adjusting your woodburning habits. To be effective, the combustor must be located directly above the stove's firebox so

**Catalytic wood burners** need to be specially designed; this diagram of the Lavec shows one approach. The catalytic combustor sits above the combustion chamber exposed to the heat and smoke. Under the operating conditions shown, air for primary combustion comes in under the stove and enters the combustion chamber at the front to feed the wood fire. Secondary combustion air enters between the double glass of the doors and is thus heated before it reaches the catalytic combustor. There, most of the flammable products in the smoke are burned. The exhaust gases rise to the secondary heating chamber, then descend to exit through the flue. When the Lavec's doors are opened,

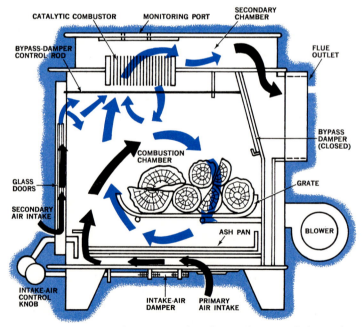

Inside look at a catalytic wood stove

as when you're starting a fire, the bypass damper also opens, forming a direct path from the combustion chamber to the flue so the exhaust gases flow directly out. (The blower can be used to boost heat circulation from the exterior surfaces.)

**Most new catalytic stoves** use screen over combustor inlet to protect it from damage and to initiate turbulence in gases. Webster uses ported "burner" as combination air inlet and guard; under some phases of firing, it had appearance of gas burner.

it "sees" the smoke when it is hottest. A bypass damper, operated either manually or automatically when the stove's door is shut, diverts smoke past the combustor until its firing temperature (500° F) is reached.

Since the catalyst doesn't operate until it reaches this temperature, you need a hot kindling fire to get it going from a cold start. You must make sure the damper is open at this time to protect the combustor from direct flame exposure. During this period the stove operates like any wood stove, but there is an excess of combustion air, a condition that tends to yield little creosote. When you have a sustained hot fire going, you close the damper and the combustor becomes active.

Knowing when to close the damper can be a matter of intuition, or you can add some auxiliary temperature gauges. But don't be tempted by the viewing port (a heat-resistant-glass window on most stoves that allows you to watch the combustor in operation) for a visual cue. The cell burns with a bright red glow, but only during the upper limits (above 1,000° F) of catalytic activity. It may also be active when it's not glowing.

A better way to judge activity is to install a thermometer (a magnetic type will do) near the viewing-port window. The temperature will rise markedly and rapidly when the combustor ignites. It's also a good idea to mount a thermostatic probe in your stovepipe to measure the temperature of flue gases. Flue temperature drops in a well-designed stove after combustor ignition because much of the heat is being transferred to the room. Such a probe will also warn of an overheated chimney or possible fire. One unit that I've looked at is Chimney Fire Alert (The VT Group, RD#1, Norrisville, VT 05661).

Once the combustor is lighted, one of the advantages of catalytic stoves will soon become evident. You get astonishingly long burn times, so you don't have to feed the stove as much. Catalytics use about one-third less wood, on average, than conventional stoves for the same heat output. I've gotten better results by loading them with the same frequency but using less wood, rather than the other way around. That avoids "dumping" the combustor by letting it cool down too much. Remember, you don't want the actual temperature to drop below about 500 degrees. If it does, the combustor will go out and you'll have to refire for a few minutes.

As with any wood stove, dry, seasoned wood works best. The higher moisture content of green wood can lower the flame temperature and consume heat that could be used for heating your home. Unlike conventional stoves, catalytics are designed to burn natural wood only. Don't use them as incinerators: Such things as

## The latest catalytic wood burners

In the newest catalytic wood burners to reach the market, the differences I've found are not spectacular but are significant.

Several of the new stoves show a lot of attention to baffling procedures to get the most from the additional generated heat. After all, the excess heat available from the combustor would simply be wasted if it were allowed to flow up the chimney. In our Webster test unit (opening photo), the ornate design hides an elaborate, deeply finned heat sink weighing nearly 100 lbs. Timberland uses a U-channel heat exchanger and a "dropped" combustor to obtain heat-transfer characteristics.

Some of the new units contain a new precision thermostatic control to produce desired heat output while maintaining combustor operation. The device is made by Condar, and I tried one of the laboratory models the company developed for testing and demonstration.

What this lab "mule" lacked in appearance it more than made up for in performance: It recorded burning times of 27 to 28 hours and exhibited excellent heat transfer. I also tried out a test version of the pedestal-mount Royale Scot adapted for the Condar control. While the heat-transfer characteristics of this stove weren't as good as the Condar lab model (the designer was still experimenting with the baffling when I received the test model), it proved to me that the thermostatic control is an effective solution to the control problems that plagued some of the early catalytic stoves.

Other new products extend the catalytic principle beyond wood stoves. Hart has introduced the first catalytic fireplace insert with combustion efficiency of up to 95 percent. And Steel King's new wood-fired boiler and warm-air furnace have fireboxes that are specially designed for catalytic combustion.

---

papers with colored inks, plywood scraps, painted or stained wood, and compressed logs with binders contain chemicals that can damage the thin metallic coating on the combustor cell. And if your stove is designed to burn coal as well as wood, you must remove the catalyst before burning coal.

After the fire is rolling along, you can adjust the air-intake dampers to maintain the best level of catalytic activity for your particular needs. You will find, though, that even with the dampers throttled down to lowest firing levels, the stove will put out a great deal of heat. So it's best not to use a catalytic stove on mild days (or a conventional stove, either, since air-starved operation increases creosote buildup).

Once you live with a catalytic for a short time, you'll find that proper operation is a snap; proper maintenance is even

**Finally, there's a catalytic fireplace insert,** Hart model 4011. When author viewed test in Hart's facility last winter, combustion efficiency was running around 95 percent. Combustor is warranted for 3 years.

easier. Yet, judging by the mail I've received, many people believe that plugging of the combustor is a serious problem.

Only once have I been able to plug a combustor, and that was using a non-standard, high-resistance cell—and using it entirely improperly (starting a fire with green, pitchy wood and closing the damper long before the cell reached proper temperature). Even then, only a small amount of smoke escaped into the room before the fire quenched itself.

If you follow recommended procedures it is doubtful that you would ever plug a catalytic combustor. Nevertheless, regular inspection and maintenance should not be ignored. Most stove makers have designed the catalytic units so they drop clear when stainless-steel pins are removed. Others are mounted in an insulated "can" that can be lifted out. When inspecting, check the entire wood-burning system, including safety bypasses in the stove and chimney. Even when performing properly, catalytic stoves can let some creosote accumulate.

---

Since the beginning of the energy crunch studies showed that setback thermostats or "clock thermostats" could save a significant amount of energy. They, too, were nothing new, but those in existence were rather crude compared to those that have recently been spawned by microcomputer chip technology. No more leveling or dirty contacts. They can give you complete control over your home's central heating and cooling systems every day in the week—and even when you're on vacation.

"Hot spots" in your home can be a good thing on a cold winter's day—if they are exactly where you want them. Usually they aren't, especially if you're taking advantage of an auxiliary source of heat such as a wood stove or passive solar heating. Any reader who has been exposed to such a system knows about the overheated room closest to the heat source while the rest of the home is freezing. We would have that problem here at Chestnut Mountain, especially in the front room and loft area, if it weren't for a few dollars spent for extra ducting and a couple of motorized dampers.

By now you may well be wondering what all this means to you and how it can help you reduce the cost of operating your home. First of all, we must concede that every home and every lifestyle is different. The key is to take the information here (and throughout this book), start with those concepts that best fit your circumstances and your existing equipment, then apply that part of the information that can benefit you.

Which brings us back to Chestnut Mountain: Details of our system can serve as an example of the application of this concept. First, we started with an energy-efficient structure—a home of "hybrid" construction engineered by Mayhill Homes (Chandler Road, Gainesville, Georgia) which is partially panelized and partially precut. Walls and ceilings are heavily insulated (R-30 overhead, R-19 walls), and have a polyethylene vapor barrier. We chose a Coleman heat pump, which was (and still is at this writing) the most efficient air-to-air model marketed at the time. But the heat pump is in reality a backup; our overall system also made use of a catalytic wood stove (selective thinning of the surrounding woods supplies more than enough fuel) and heat gain from solar radiation, since the home is oriented due south and the only significant glass area is on the front. Making use of what we have, remember?

Keeping tabs on all this is the Smartstat electronic thermostat, plus a circuit that has additional sensors located near the peak of the 22-foot ceiling that sense heat gain from the wood stove or solar radiation and give priority to that heat. When temperature at the ceiling is near 80 degrees, the thermostat turns on a blower in the furnace and closes dampers in the return air ducts so the majority of air flowing through the furnace is coming from the ceiling area. This keeps the loft area cool and circulates filtered, humidified, low-level-heated air thoughout the lower levels of the house.

When temperatures are mild, I let the heat pump carry the heating load, because it is highly efficient down into the lower 40's. (On a clear day it doesn't run, because solar gain is ample to maintain comfortable levels down to about 18 degrees outdoor temperature, even without storage other than the mass of the house.) When temperatures reach down into the 30's, I throw a few sticks of wood into the stove, taking advantage of it when it's most efficient—in colder weather.

The important thing to understand here is that this concept works—my energy consumption at Chestnut Mountain is only about 25 percent of that required for a typical home of its size (about 2400 square feet). Even more important still, the proper application of the right ideas and equipment can work just as well for you.

# Programmed heating and cooling improves your comfort, lowers your energy bills

*Microprocessor controls in window air conditioners and whole-house heating-cooling systems give new flexibility*

## By Evan Powell

Microcomputer-based controls have added versatility, reliability, and accuracy to many new appliances and tools. Now computerized air-conditioning and heating systems can bring additional benefits: greater comfort, energy economy, and easier servicing.

Computerized air-conditioning controls automatically adjust temperatures upward at preset intervals for greater energy efficiency. You needn't waste electricity cooling rooms when no one is around. And recently, testing a new Sears computerized window model, I discovered some fringe benefits: A built-in computer can help prevent compressor overheating and damage following a power failure, and it can shut off the unit during low-voltage brownout conditions.

Microcomputers are also being used with heat pumps. And a number of companies are marketing retrofit computerized controls for existing air-conditioning and heating systems.

## TEMPERATURE ECONOMY

The versatility of computer programs ensures the most economical operation from your system. The value of temperature set-back for heating (and set-up for cooling) has been established in several studies.

A "real-life" two-year study sponsored by Honeywell at Edmond, Oklahoma, verified previous computer simulations. Honeywell also found that, for cooling, homeowners saved most when they set their thermostats up by five degrees or more for 17 hours per day.

"It's very important to set up during the right part of the day," says Cliff Moulton, manager of marketing for Honeywell. The ideal set-up time for most homes in the Edmond study was 10 a.m. to five p.m.—hours when family members are likely to be out. The families whose

21

homes were monitored by sensing devices and recording equipment saw energy bills drop an average of 23.4 percent from May to September.

Comfort is another benefit with computerized control. While testing a General Electric control for heat pumps, GE's manager of applications engineering, Chuck Erlinson, installed the microprocessor unit in his home with special circuits so the existing conventional thermostat could be switched in and out. Chart tracings of the temperature made during the monitoring period were much more constant for the microprocessor unit than for those with the conventional thermostat.

This corresponds to my findings at our experimental studio home with a SmartStat 1000 coupled to a Coleman high-efficiency heat pump. The SmartStat, marketed by Johnson Controls, is a programmable set-back/set-up control for heat pumps that replaces an existing thermostat.

Reliability is another reason for using computerized controls. I've found that any electronic device that's going to fail will usually fail within the first few weeks of operation. That's covered by the warranty. After that, there's none of the mechanical wear and arcing that occurs in conventional controls and reduces their operational lifetimes.

## COMPUTERIZED ROOM COOLING

Sear's model 17905, a highly efficient (9.6 EER) 9,000-Btu model, is the first window air conditioner to use computer-control technology. I discovered its versatility during the pleasantly warm but humid days of early spring in South Carolina. Tap the HUMIDITY CONTROL pad on the touch panel and the unit cycles on at intervals to remove moisture with minimum cooling. According to Sears, it's the equivalent of a 20-pint-per-day dehumidifier. In the cooling mode, the control gradually slows the fan as the desired temperature is reached, keeping power consumption and noise down. Electronic temperature sensing keeps temperatures very consistent.

Short-cycle protection is another feature built into the Sears unit. In conventional models, a brief line-voltage interruption can damage windings when the compressor attempts to restart before pressures equalize and unlock the rotor. The Sears model prevents this by delaying start-up for three minutes after power returns. You have to reprogram the computer after a power interruption.

Another protective bonus in the Sears microprocessor control is built-in brownout protection. If the voltage falls below a predetermined level, the unit shuts off until the normal power level is restored.

But the microprocessor's most important job is controlling set-up periods when no cooling is required. In the Sears unit, this takes the form of an on-off cycle. You set the shutdown time by touching the OFF TIME pad, then programming the time you'll leave home. Next, touch ON TIME and enter a time about one hour before your return. This program stays in the unit's electronic memory until it is changed or erased. On weekends or at other times you can override it by touching the TIMER ON/OFF pad.

The NIGHT SETTING pad and its program allow the temperature to rise four degrees one hour after you touch it and go to bed. Five hours later, before wake-up time, the thermostat returns to the original cooling setting.

The high-efficiency (9.6 EER) 9,000-Btu Sears model sells for about $550.

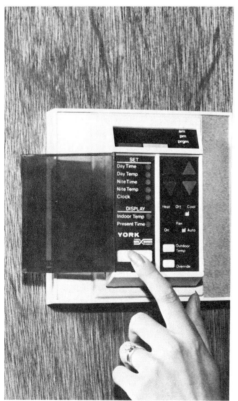

**Front panel** of computerized air conditioner from Sears (top left) slides open to reveal sophisticated controls. Touch pads (top right) let you enter night setting, start-stop times, humidity control. York's thermostat for its Enmod heat pump (left) also lets you program automatic night set-ups for energy savings. GE's DMC 2000 control (above) indicates time of day and indoor or outdoor temperatures.

## CENTRAL-COOLING CONTROLS

I've already mentioned the SmartStat, the first set-back/set-up control intended for heat-pump systems. Now there's a new kid in town. General Electric's DMC 2000 operates only with its new Executive II heat pump, which features a two-speed compressor. The 2000 has a self-diagnostic feature, which monitors sensors within the system. If anything gets off course, a service light alerts the user to notify the repair department. The service technician plugs in an analyzer, and the control indicates the trouble area.

The DMC 2000 provides up to two temperature changes a day, and its display can also indicate outside temperature. When remote sensors are used, it averages the input for an indication of temperatures throughout the house. The STANDBY pad keeps temperatures in an unoccupied home from rising above 85 degrees F or falling below 55 degrees.

York's computerized control, used in conjunction with a Yorkguard computer module in its heat pump, also provides a self-diagnostic feature. Electronic sensors control defrost time and auxiliary-heat input based on actual outdoor conditions.

York has also put microcomputer technology to work in its Enmod residential heat pump. In this model, which costs about twice as much as York heat

pumps with single-speed compressors, the speeds of both the compressor and fan are varied. Instead of cycling the compressor on and off, the Enmod modulates the 60-Hz line current to lower frequencies, using a new inverter and microcomputer control to vary compressor and fan speeds. The company claims Enmod consumes 17 percent less energy than its conventional heat pumps.

## RETROFIT CONTROLS

Both Honeywell and Texas Instruments are marketing retrofit computerized controls for air-conditioning and heating systems, but not for heat pumps at this time. TI's 4FA23 has three independent operating modes that give a total capability of seven different temperatures and six different times. It wires directly into the existing 24-volt thermostat circuit in most cases, although oil furnaces and gas furnaces with electronic ignition may require the addition of an isolation relay.

Honeywell's T-800 control, also for conventional heating-cooling systems, allows for day and night set-back periods. A FIVE-DAY/SEVEN-DAY selector switch allows you to skip the day set-back period on weekends.

All of these central controls have self-contained batteries to prevent loss of programs during brief power outages.

# CENTRAL HEATING— OIL, GAS, OR HEAT PUMPS

**W**hat's the best heating system of all? There's no pat answer for that, for there are too many variables, such as the cost and availability of fuel at any given time. And unless you're building a new home, much will depend on the resources you already have in your present system, and on its age and condition.

One thing is certain, however—it pays to know as much as you can about the options available to you in the equipment that creates the heat in your home. You should know how the different systems work and understand the advantages and disadvantages of each type of system, as well as the requirements for making use of them. The material that follows in this chapter is the starting point, whether you hope to upgrade your present equipment or are planning a complete new installation.

# High-efficiency heating systems help bring winter fuel costs back to earth

## Part 1: Oil-fired systems

*First-rate heating systems that can save you money are available now. Even better ones are coming*

## By V. Elaine Smay

For every dollar you spend on fuel to heat your home, the equivalent of 35 to 60 cents is probably going up your chimney. But it doesn't have to.

Conventional heating systems now on the market can cut your fuel waste significantly, especially if your present system is on the low end of the efficiency scale. Furthermore, new-technology systems *may* cut that waste even more—if they can be successfully brought to market.

But before I give you the details on conventional and advanced oil-burning systems and gas-fired equipment, you need to understand why many existing systems are so inefficient, and why rating efficiency isn't as simple as it might seem.

Why aren't all furnaces and boilers designed to extract every possible Btu from each gallon of fuel they burn? Why do some of them waste an astonishing 60 percent of the fuel that goes into them? The reason can be mainly traced back to the fact that they were designed when a gallon of oil cost 15¢ and a therm of gas about the same amount. "There was no incentive to make them efficient," says Dr. James Drewry, manager of residential and commercial utilization at the Gas Research Institute in Chicago. "The only goal was to make them cheap."

In addition, many home-heating plants are oversized: They burn fuel at a higher rate than is necessary to meet the heat load of the house, and the excess heat just wafts up the chimney. This is especially true of boilers that contain coils to supply the domestic hot water without benefit of a storage tank. "You have to be able to satisfy the guy who takes a 20-minute shower," says John Batey, former head of a home-heating equipment testing program at Brookhaven National Laboratory. And that may mean a much higher firing rate than the house requires. While the Brookhaven testing program has concluded, advanced research in home heating continues at Oak Ridge National Laboratories and several private and government-sponsored research facilities.

**At Brookhaven National Laboratory,** John Batey (left), former director of the heating-system test laboratory, and Roger McDonald (right) monitor computers as they give instantaneous readouts of the performance of the oil-fired boilers under test. In the background, Tony Kokinelis takes a sample of exhaust gases for infra-red analysis of one boiler under test.

Furthermore, when you insulate and weatherize your house you reduce the heating load and therefore exacerbate the oversizing problem. "You've reduced your total fuel use, but you're using it even less efficiently," Batey points out.

## MEASURING EFFICIENCY

Burn a gallon of fuel oil and you release 138,800 Btu of heat. In your furnace or boiler, some of those Btu are extracted by the heat exchanger and used to heat the air or water that goes to heat your house. The rest are lost—either up the chimney or through the jacket of the heater, for the most part. Ideally, the rated efficiency of a heating system would tell you what percentage of the Btu that go in will actually contribute to heating your house. In practice, however, it's not that simple. Read advertisements and catalogs and you see terms like combustion efficiency, seasonal efficiency, overall efficiency, steady-state efficiency, and on and on. There are almost as many kinds of efficiency ratings as there are heating systems. And there are different measurement tests to determine them. Without knowing what the ratings mean or how they are determined, you can't make intelligent use of any of them.

Two government programs of the late 70's helped eliminate some of the confusion. One was the heating-equipment testing program, set up by the Department of Energy (DOE) at Brookhaven National Laboratory. The other is an efficiency-labeling program mandated by the Federal Trade Commission (FTC). Both programs sought to predict what is called the steady-state and the seasonal efficiency of a heating system. The rub is they used different tests to do it.

The Brookhaven testing lab rated the efficiency of a system based on direct measurements. This required frequently analyzing the oil to determine its precise Btu content, then measuring the amount of heat that was actually transferred to the boiler water. The difference between input and output Btu was wasted heat.

In the lab at Brookhaven, the boilers under test were equipped with sensors that measured temperatures, flow rates, and other critical variables. The sensors fed their data to computers that whirred away, printing out the results and flashing instantaneous readings via red LED displays.

To determine the steady-state efficiency of a system, the burner was fired up and allowed to run for a specified time period; then data were collected. The steady-state efficiency was the maximum efficiency of a heating plant.

A second sequence of tests was used to arrive at a seasonal-efficiency rating. That's the number that is supposed to predict the system's actual performance in a given house throughout the heating season. Except for high-technology equipment such as the pulse combustion furnace, the seasonal efficiency will always be lower than the steady-state efficiency because, in real life, the burner is only on a small percentage of the time—from 15

to 20 percent is common. And while the system is off, it continues to lose heat. The seasonal efficiency must take into account these standby losses. The less the burner is on, the greater the standby losses.

To determine a heating system's seasonal efficiency, the Brookhaven tests cycled each system on and off, varying the length of the cycles from on all the time to total shutdown. Then, using a computer program that included a given heat load for a theoretical house and hourly weather data averaged over 10 years, the seasonal efficiency of a system was predicted.

The FTC labeling program, on the other hand, bases its efficiency ratings on an indirect testing method. Technicians measure the temperature of the flue gases and the proportion of carbon dioxide in them (the more $CO_2$, the higher the efficiency). This method assumes all other Btu were turned into useful heat. The steady-state efficiency is based on that assumption. Then, to arrive at a seasonal efficiency, other standard assumptions are made to account for standby losses.

The Brookhaven engineers felt their direct method to be significantly more accurate than the FTC's indirect tests. The difference in results will usually vary from zero to 12 percent, they've found. In a few cases, it's even greater. "The FTC's method tends to make poorer boilers look better," says John Batey, "so it gives little incentive to buy the best equipment."

But if the labeling program isn't perhaps as accurate as it could be, it should at least standardize procedures and make comparable numbers available to the consumer.

## THE BROOKHAVEN TESTS

In a barracks-like building near the edge of the sprawling Brookhaven campus on Long Island, New York, both commercially available heating systems and new-

## Modifications and add-ons—how much can they help?

One aspect of the Brookhaven program has been to determine the cost-effectiveness of various changes you can undertake, ranging from simply having the burner adjusted to installing a whole new heating system. Some add-on items also have been evaluated. Of the actions listed below, items 1, 4, 6, 8, 9, 11, 12, 13, 15, and 16 were actually tested at Brookhaven.

The other evaluations are based on tests by other organizations and published reports. The payback period is based on 1500 gallons of fuel per year at 50¢ per gallon. Since you're probably paying more than that—perhaps twice as much—payback periods would be shorter; but costs of some actions may also have risen since this chart was compiled.

| Action taken | Est. fuel savings (%) | Approx. cost ($) | Payback period (yrs.) |
|---|---|---|---|
| (1) Replace burner with high-speed retention-head type[a] | 16[b] | 250 | 2.1 |
| (2) Thermostat setback[c]—automatic | 8 | 80 | 1.3 |
| (3) Thermostat setback[c]—manual | 8 | 0 | 0 |
| (4) Boiler water-temperature reduction (by 35° F) | 5 | 0[d]-20 | 0-0.5 |
| (5) Burner-efficiency adjustment | 3 | 0[e]-30 | 0-1.3 |
| (6) Reduce burner firing rate (by 25%) | 8[b] | 0[e]-25 | 0-0.4 |
| (7) Put in boiler fire-tube turbulator | 5[f] | 50 | 1.3 |
| (8) Install vent damper[g] | 10[b] | 200 | 2.7 |
| (9) Install stack-heat reclaimer[g] (economizer) | 15[b] | 350 | 3.1 |
| (10) Duct combustion air from outdoors[h] | 0-3[i] | 100 | 4.4 |
| (11) Change to modern high-efficiency burner-boiler | 24[b] | 1500 | 8.3 |
| (12) Change to blue-flame burner-boiler | 24[b] | 1500 | 8.3 |
| (13) Change to low/variable-firing-rate burner[j] | 20 | 500 | 3.3 |
| (14) Outdoor boiler installation | 0-10[k] | — | — |
| (15) Combustion-air humidification | 1 | 200 | 26 |
| (16) Burn water/fuel-oil emulsion | 0 | — | — |

**FOOTNOTES:** (a) Firing-rate reduction should accompany burner installation. (b) Based on dry-base steel boiler with conventional burner. (c) Setback 10° F for eight hrs. a day. (d) Manual adjustment by homeowner. (e) May be included as part of annual servicing. (f) Applicable only where turbulator was removed from boiler during cleaning. (g) Possible safety hazard—long-term testing required. (h) Including inlet damper for burner off-cycle. (i) Will vary depending on boiler location in structure. (j) Commercial equipment not yet available. (k) Will vary with boiler—testing required.

technology prototypes were put through those rigorous efficiency tests. When I visited the laboratory in 1979, it was only set up to test oil-fired boilers; but gas-fired systems and forced-air furnaces were tested, also. (In the heating industry, a warm-air system is called a furnace; a hot-water or steam system is called a boiler.) The Brookhaven team also tested various retrofit options and add-on items to determine whether they improve the efficiency of a heating system. The table gives its conclusions.

"Of the equipment that's commercially available right now," says John Batey, "we've found a 30-percent difference in annual fuel use. The best will give a seasonal efficiency of around 76 percent, the worst around 53 percent."

Another finding about currently available equipment: High-efficiency units seem to share certain characteristics. Identifying such characteristics—if they existed—was a primary goal of the program. Two principle ones emerged.

The first concerns the burner type. Brookhaven researchers said that most high-efficiency systems have what is called a high-speed flame-retention-head burner. The high-speed designation comes from the fact that the motor operates at a higher rpm than that of a conventional burner. Retention-head refers to its tendency to burn with a flame that stabilizes near the head. "It's really not extraordinarily different from a conventional burner," says Batey, "except for the head where oil and air mix. Air and fuel are mixed better, so less excess air is required to produce a low soot level. Thus, you increase efficiency."

What Batey refers to is this: Under ideal (stoichiometric) conditions, only the amount of air needed for combustion would be fed into the combustion chamber. In reality, however, all burners must burn with some excess air to keep smoke and soot formation down. But the more excess air you have, the more it tends to remove heat from the system and reduce efficiency. (Also, the more excess air, the less $CO_2$ in your flue gasses; that's why measuring $CO_2$ content, as the FTC test method does, is a reasonable indicator of efficiency.)

Not only does the high-speed flame-retention-head burner burn with less excess air, it also tends to reduce standby losses. In field tests where conventional burners were replaced with retention-head types, the average saving was around 15 percent.

The second feature that the Brookhaven tests showed to be important to the efficiency of a boiler was the design of the heat exchanger. Cheaper boilers will usually have a dry-base heat exchanger: Hot gases pass upward from the insulated combustion chamber, into tubes immersed in the water, and from there up the stack. The heat must be transferred from these tubes to the water. But the type that proved to be more efficient is a wet-base heat exchanger. With this type, the water jacket actually drops down along the sides of the combustion chamber as well as through the heat-exchanger tubes. "We don't want to say that all wet-base boilers are better than all dry-base boilers," cautions Batey, "but in general, it's true."

## BLUE-FLAME TECHNOLOGY

The heating-system test lab at Brookhaven was actually part of a broader program that also included the development of new equipment. This aspect of the program channeled DOE money to manufacturers working on promising new technologies. The first commercially available equipment to come out of this was a boiler using blue-flame technology. In a blue-flame burner, a baffle system recirculates

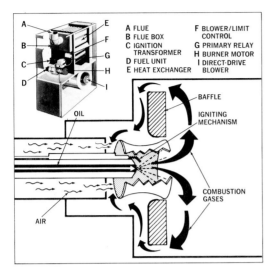

A FLUE
B FLUE BOX
C IGNITION TRANSFORMER
D FUEL UNIT
E HEAT EXCHANGER
F BLOWER/LIMIT CONTROL
G PRIMARY RELAY
H BURNER MOTOR
I DIRECT-DRIVE BLOWER

OIL

AIR

BAFFLE

IGNITING MECHANISM

COMBUSTION GASES

**Blueray oil burner,** developed partially under a Department of Energy contract, uses a special baffling system to recirculate combustion gases to preheat the incoming oil. The result is a flame that burns hotter—it's blue instead of the typical yellow oil flame. Blueray furnaces (above left) have been on the market for several years; two boiler models are also available.

combustion gases (see diagram). This preheats the oil prior to combustion and results in a hotter flame that burns blue, like a gas flame, rather than yellow, as does a normal oil flame. The hotter flame means more efficient burning with less soot—at least in theory.

Blue-flame technology actually got its start in the 1960's when a group of small oil dealers set out to improve the oil burner to better compete with natural gas. For years the technology languished. But in 1977, the Meenan Oil Co. of Syosset, New York, through a division called Blueray, began marketing two blue-flame furnaces. Under the Brookhaven contract, Blueray developed boilers of six different sizes and one new furnace.

A prototype of a Blueray boiler completed seasonal efficiency tests at Brookhaven with good, but unspectacular, results. "It was one of the highest efficiency units we tested," reported Dick Krajewski, head of the test lab. "But no better than the best conventional systems."

"Our big advantage," responded Joseph Incrocci, president of Blueray, "is that we do not generate soot; therefore our efficiency doesn't decline as soot builds up."

That's a controversial point. Brookhaven technicians inspected Blueray furnaces that had been in homes for one or two years for soot accumulation. They had intended to look at 25, but they found the first 12 clean, so they called off the inspection. "I can show you Blueray furnaces that have been in the field for five years and you'll see no soot," says Incrocci.

## DEVELOPING TECHNOLOGIES

While current technology produces very good heating equipment, various new systems are pushing to do even better. The many ways that engineers are seeking higher efficiencies in oil-fired systems can be divided into two general approaches: lowering firing rates and condensing flue gases.

### Lowering Firing Rates

Conventional oil burners can't operate with an oil-flow rate less than about a half gallon per hour. Below that, the size of the nozzle is too small; it tends to plug and cause maintenance problems. But some systems under development are designed to reduce the firing rate. This would lower the temperature of the stack gases and therefore reduce the amount of heat wasted up the chimney. A Brookhaven study measured stack temperatures in 100 homes and found that the average was 700 degrees F. Sometimes temperatures as high as 900 degrees were encountered. Good equipment today operates with a stack temperature of 400 to 500 degrees. But it could go lower. A lower firing rate also means the burner is on a larger part of the time. Thus standby losses are reduced.

Once an oil burner can fire at a very low rate, the next step toward maximizing efficiency is to vary the oil-flow rate. By using electronic controls you can modulate the flame size according to the changing needs of the house. Traditional heating systems have two modes of operation: They're either on or off. But with a modulating system, when temperatures outside are very cold you might be burning a gallon of oil an hour. Then as the weather warmed up, the firing rate would decrease to a fraction of that. By constantly matching the heating needs of the house you decrease the time the burner is off. Ideally, you would have a system that was on all the time; then it would actually operate at steady-state efficiency.

One approach to a low- and variable-firing-rate burner is being explored by inventor Robert Babington. Babington invented a unique system of atomization whereby the liquid to be atomized flows over the surface of a sphere, while air is forced out through a slot in the sphere (see diagram). The airstream breaks up the liquid flowing over the slot into minute particles. The result is a superfine mist that has been used for many years in medical applications. In an oil burner, the Babington system of atomization should result in very even burning and very fine droplets, which means it could operate with little excess air. And since no liquid flows through the slot, no clogging can occur. Babington's burner also offers the possibility of operating on just about any liquid fuel.

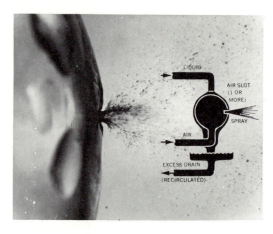

**The Babington principle of atomization** is illustrated in the diagram above. The liquid to be atomized (in this case, the fuel oil) flows over the outside of a sphere. Air is blown out a slot in the sphere. As the flowing liquid reaches the slot, the airstream breaks it into a fine mist (photo). (This represents the concept, not the configuration of Babington's present burner.)

Babington first attempted to turn his principle into an oil burner by licensing its development to Rocketdyne. But for a number of years he's been pursuing it himself.

A feasibility model of the burner was tested at Brookhaven. The report concluded that it could operate efficiently down to 0.2 gallon per hour. The prototype could not be put through cycling tests, however, so a seasonal efficiency could not be determined.

Another path to lower firing rates and modulated heat output is through ultrasonic atomization. Sono Tek's ultrasonic nozzle has an acoustical resonator that is driven at frequencies from 60 to 80 kHz. These high-frequency sound waves break up the fuel into microscopically small droplets. The manufacturer claims it can be fired down to 0.1 gph. A prototype was put through the steady-state efficiency tests at Brookhaven and tested at firing rates between 0.3 and 0.5 gph. "The Sono Tek successfully demonstrated the feasibility of ultrasonic atomization for efficient residential heating," concluded the Brookhaven report. But the prototype burner required a complicated manual startup procedure and therefore could not be put through the cycling tests to determine seasonal efficiency.

## Condensing Flue Gases

Low and variable firing rates can theoretically increase steady-state efficiency to around the 90-percent area. But to wring the remaining Btu out of the fuel oil, you have to cool the flue gases below the dew point and capture the latent heat of vaporization (the heat released when a gas condenses).

Traditionally, heating equipment has been designed to *avoid* condensation in the heat exchanger and flue. Such condensate contains corrosive and caustic products that can cause many problems. But now, some systems are designed to deal with the condensate to reap the efficiency benefits. An additional bonus: Once you cool the flue gases to 200 degrees F or below, you may not need a chimney; flue gases can exit through a hole in the side of the house.

One such condensing system I saw working was invented by Borje Ollson of Kivik, Sweden, and is being manufactured by a Swiss company. The unit, called the Turbo Puls, uses a pulse combustion principle, not unlike the engine in your car (see diagram, page 34). A closed system, it draws combustion air through a pipe from outside and exhaust gases exit through an insulated, 25-to-30-mm polyvinyl chloride (PVC) tube. The unit is quiet and compact; it can burn either oil or gas.

In tests at Brookhaven, a prototype Turbo Puls boiler measured around 90 percent steady-state efficiency. But during cycling tests it produced condensate in the combustion chamber and failed, prompting a redesign of the problem area.

While these new-technology oil burners sound very promising, bringing them to market is not easy. As John Batey says, "You meet the law of diminishing returns: Going from 55- to 85-percent efficiency is a lot easier than going above 85."

And Robert Babington, after working many years to commercialize his oil burner, talks about other realities of the market place. "When you see that the efficiency is good, you have to look at your burner and say: How much will it cost? How long will the thing last? Will the oil dealer be able to service it?"

Those are tough questions. But the incentive to find positive answers is certainly at hand. Supporting an oil guzzler at $1.25 or more per gallon is pretty tough, too.

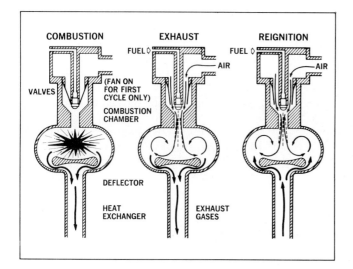

**Pulse combustion** is a new technology for home heating. Above is a schematic of how the Turbo Puls boiler works: (1) A small amount of fuel and air are sent into the neck above the combustion chamber, where they mix. At the combustion chamber, a spark ignites the mixture. A small explosion occurs. (2) The hot gases swirl around in the combustion chamber and out the exhaust; heat is transferred to the surrounding water jacket. (3) As gases rush out, pressure in the combustion chamber drops below ambient. This opens the valves, letting in more air and another drop of fuel. Hot exhaust gases also rush back into the combustion chamber and ignite the mixture. (A spark is needed only for the initial explosion.) Explosions take place at the rate of 87 per second. The Turbo Puls is designed to condense the water vapor in the flue gases, thus capturing additional heat. At right, inventor Börje Ollson stands beside a prototype of his boiler.

# High-efficiency home heating

## Part 2: Gas-fired systems

*Your future space heater may be mildly—or wildly—different, but it won't be a gas guzzler*

## By V. Elaine Smay

Some 42.9 million American homes are heated with natural gas, according to the American Gas Association (AGA). And those furnaces and boilers operate at an efficiency somewhere between 40 and 65 percent, many wasting more than half of the energy in the costly gas they burn.

Thanks to the Federal Trade Commission's Energy Guide labels, such gas guzzlers don't sell well anymore. The labels' posted efficiency ratings are based on standard test procedures formulated by the Department of Energy, and designed to tell you how much useful heat that system will extract from the fuel it burns over the course of a heating season.

While existing gas furnaces and boilers (a warm-air system uses a furnace; a hot-water or steam system uses a boiler) average only around 55 percent seasonal efficiency, the *best* systems on the market, using conventional technology, operate at around 70 percent. With this season's new commercially viable products, that number has risen by a substantial margin. Indeed, with the Lennox Pulse and new recuperative gas furnaces, seasonal efficiencies above 90 percent are a fact. Other R&D efforts are working toward the development of gas-fired heat pumps. If successful, this could result in dramatically higher efficiencies.

## TODAY'S EQUIPMENT

The best conventional systems available today, the ones that run with a seasonal efficiency of around 70 percent, have two things in common: an electronic ignition instead of a standing pilot, and a vent damper to minimize heat loss up the chimney when the burner is off.

A pilot consumes about 6000 cubic feet of gas a year while the burner is off. Eliminate that and you increase efficiency by four to six percentage points annually, according to Dr. Robert Wilson of Arthur D. Little, Inc. The vent damper offers even more potential for savings—from eight to ten percent, concluded a study by the Little group. Virtually every major manufacturer of gas heating systems offers units with electronic ignition and a vent damper.

## LOW-VOLUME BOILERS

When the burner in a boiler comes on, it must heat up the firebox, the heat exchanger, and the water. In older boilers especially, these constitute a large thermal mass that is slow to heat up—and slow to cool down. Once the burner goes off, the water pump also stops with most control systems. Thus the heat that remains in the boiler is dissipated out the jacket (a loss especially if your boiler is situated in an unheated space) and up the chimney (a loss wherever it's located).

In newer boilers, materials with less thermal mass are used to lessen this effect. Now the trend has been taken a step further by using only a small quantity of water in the boiler.

Evan Powell has written about one of these systems, the Paloma Pak, a Japanese import. The Paloma Pak holds only a quart of water. The distributor, NEGEA Energy Products in Worcester, Massachusetts, claims energy savings on the order of 20 percent, but has not run controlled tests.

From England comes the Potterton, a cast-iron low-volume unit, also available from NEGEA and Southern California Gas Company in Los Angeles. It promises even more energy savings because it uses electronic ignition and needs no chimney. Combustion air is drawn in by a fan from outdoors; the fan also forces the exhaust gases out a small vent tube.

An American-made low-volume boiler from Teledyne-Laars, a California company, adds another energy-saving feature: a modulating burner. This can save energy because, as the temperature outside goes up and down, it can burn at a higher or lower rate to match the heat needs of the

house. A conventional burner can only be on or off; the more it's on, the less it is subject to standby losses. Theoretically, if a burner could modulate over a wide enough range, it would run throughout the heating season and operate at the highest possible efficiency all the time.

"With a vent damper and electronic ignition, our boiler operates at a seasonal

**Very-low-volume** gas boiler, the Paloma Pak, hangs on wall. When hooked to a separate storage tank it can also supply domestic hot water. Tests in England indicate such systems can save 10 to 15 percent.

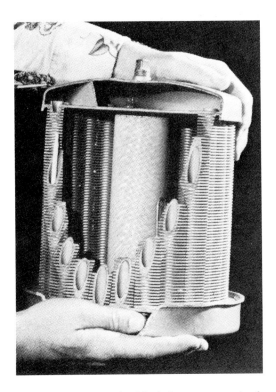

**Amana Heat Transfer Module,** a compact, efficient heat exchanger with powered gas burner, is part of a unit that also provides electric air conditioning. It is installed outside; an ethylene glycol solution transfers heat to the house. New recuperative model incorporates system into single condensing furnace, heats both air and domestic hot water.

efficiency of 76.2 percent based on the DOE tests," reports Jack Sargent of Teledyne-Laars. "And it remains within three percentage points even at the lowest firing rates."

## HEAT TRANSFER MODULE

Another unusual gas heating system now on the market is Amana's Heat Transfer Module (HTM), developed from space

technology by its parent company, Raytheon. The furnace is small and efficient, with a fin and tube heat exchanger now replacing its earlier ball matrix design. (Amana combines it with an electric air conditioner and sells the complete system only.)

In the center of the heat exchanger is a powered burner. It's called that because a blower supplies its combustion air. A standard gas burner is open to the atmosphere. In a powered burner the blower draws in the proper amount of air to mix with the gas. That means the system can burn with less excess air than is normally required for complete combustion. Efficiency is improved because excess air tends to cool down the system.

"Our HTM, depending on size, will operate with a seasonal efficiency of 81½ to 84 percent," says Chuck Muller, an Amana engineer. In its newest version, with a condensing section, it is expected to reach well into the 90's.

Traditionally, boilers and furnaces have been designed to send a lot of heat up the chimney deliberately. Flue gases contain water vapor, among other things, and the object was to keep it from condensing. If temperatures fall below the dew point, a corrosive brew results that can damage both heat exchanger and the flue. When fuel was cheap, you could throw away Btu to skirt the problem.

## CONDENSING SYSTEMS

Now the picture is different. So new systems are being designed to wring every possible Btu out of the fuel by cooling the flue gases below the dew point, which brings a sizable benefit. As the vapor condenses, it releases large quantities of heat called the latent heat of vaporization. With a gas heating system, that can mean a nine percent improvement in efficiency. Of course, it creates problems, too, in

**Variable-firing-rate,** powered gas burner developed by Foster-Miller, modulates over wide firing range—10,000 to 100,000 Btuh for this model. Here's how it works: The blower draws air into inlet manifold; "zero regulator," also called a negative-pressure regulator, provides gas to the manifold. Air/gas valve keeps the same fuel/air ratio regardless of setting, and burns with about 15 percent excess air (compared with about 40 percent for conventional burner). The gas/air mixture passes through flame holder, where an electric spark lights the flame. Combustion is complete within two inches of the flame holder, so firebox can be compact. Powered gas burners are similar to burners used in oil-fired systems, and might be used to adapt an oil unit to burn gas.

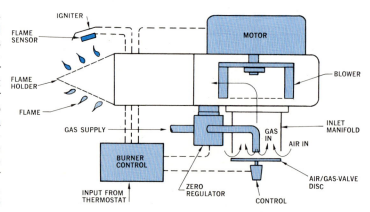

dealing with the corrosive liquid that results. So new materials, mostly plastics and ceramics, are being examined for possible use in corrosion-resistant heat exchangers.

Getting rid of combustion gases, even without a strong draft, is no problem. The new-technology burners force them out of the system, regardless of temperature. In some designs, so much heat is extracted from the gases that no chimney is needed; the now-warm gases can be vented through a plastic tube and routed out through a hole in the wall.

A powered burner, similar in concept to that used in Amana's HTM, is one possible approach (see the diagram of the Foster-Miller prototype). A powered burner could be paired with a high-efficiency heat exchanger that would cool the exhaust gases below the dew point. One such design (see the diagram of Therma-core system) uses heat pipes, those super-efficient heat-transfer devices that have been used for everything from cooking a roast to keeping the permafrost cool below the Alaskan pipeline.

But most condensing systems now available use a pulse-combustion burner. A pulse-combustion oil-fired boiler, the Turbo Puls, has been marketed by a Swiss company, and a gas-fired boiler, the Hydropulse, made by Hydrotherm in Northvale, New Jersey, is already available, as is the Lennox Pulse.

In pulse combustion, a small charge of fuel and air is ignited by a spark plug in the combustion chamber. A small explo-

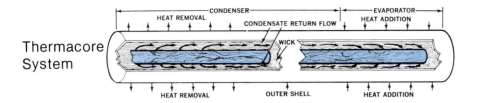

Thermacore System

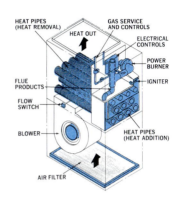

**Diagram of furnace heat exchanger** using heat pipes (above) shows heat-pipe principle: Heat input evaporates wick fluid (in this case, water), driving it to the condenser end of the heat pipe. Cooler inner walls there condense vapor, releasing stored heat. Condensed liquid returns to evaporator end by wick's capillary action and cycle repeats. Enormous amount of heat (the latent heat of vaporization) is absorbed and released as fluid vaporizes and condenses. In the furnace (right), heat from a powered burner travels around baffles, heating one end of heat pipes and vaporizing wick fluid. Vapor travels to other end where it condenses, giving up stored heat. Blower sends released heat into house ducts.

sion takes place and combustion products are forced out into the heat exchanger/exhaust pipe, leaving a low-pressure condition in the combustion chamber. This opens up the valves to admit a fresh charge of air and gas. Hot products remaining from the first pulse ignite the mix. A spark and a blower are necessary only on the first pulse.

Pulse combustion is actually an elderly technology, used most notoriously in the German V-1 buzz bombs during World War II. In the mid-'50's a Canadian company first applied it to a heating system, a boiler called the Pulsamatic. "Their selling price was about three to four times that of conventional gas boilers," says John Penny of Hydrotherm, "and when gas was cheap, nobody cared that they would save 30 to 40 percent on fuel."

Pulse combustion is a natural for condensing heating systems because the pulses force the exhaust gases out of the exhaust automatically. With a powered burner, the fan must run constantly, and that takes electricity.

Pulse combustion has another advantage over a standard powered burner: The exiting gases are very turbulent. "The hot gases take the form of a rolling doughnut," explains Conrad Yankee, whose company, Yankee Engineering, has been working on a pulse-combustion boiler. "These rolling doughnuts progress down the pipe, scrubbing the sides and transferring a lot of heat in the process—almost twice as much as an ordinary burner." Thus the heat exchanger can be smaller and cheaper and still cool the gases to below the dew point.

The Hydropulse operates at a steady-state efficiency of 91 to 94 percent, according to the company. And since it's a sealed combustion system with minimum standby losses, it's seasonal efficiency is not much lower, the company claims.

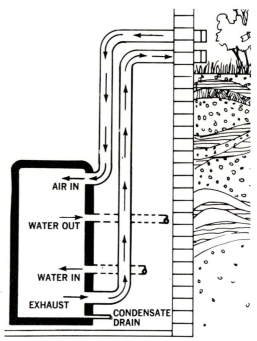

**Pulse-combustion gas boiler,** the Hydropulse, is now available in 100,000 Btuh model. Five other residential-size models are planned. Pulse-combustion systems can cool exhaust gases below the dew point, thus improving the systems' efficiency by capturing the latent heat of vaporization. Because flue gases are cool, they can be vented, via a small plastic pipe, through a hole in the house wall.

## PULSE-COMBUSTION R&D

Making a commercially viable pulse-combustion furnace is trickier than making a boiler. The problem: noise. The same noise that gave the buzz bombs their name also occurs in a pulse-combustion heating system. In a boiler, the surrounding water jacket tends to muffle the sound, but there's nothing to hide it in a furnace. The AGA Labs/Lennox unit uses a muffler to resolve the problem.

Another pulse-combustion boiler, quite different from the Hydropulse, is being developed by Yankee Engineering. While a pulse-combustion boiler is at the heart of the device, it also includes a solar loop that can be actuated with an optional control and pump module. The basic pulse-combustion/solar boiler would produce space heat and domestic hot water in a compact, inexpensive package that would be easy to install and service. The system is modular, and designed to incorporate many new technologies as they become available. Company tests of a prototype unit have shown seasonal efficiencies in excess of 90 percent without the solar module.

## GAS-FIRED HEAT PUMPS

Once you've squeezed out most of the Btu in the gas you burn by capturing the latent heat of vaporization, it takes what may appear to be a thermo-dynamic sleight-of-hand to get more heat from the fuel. The proper genie for this job is, of course, the heat pump.

A heat pump is essentially an air conditioner that is reversible. It uses mechanical energy to move heat from a cooler to a warmer location, the reverse of what happens in nature. Thus, in summer, it can extract heat from your house and reject it to the warmer outside air. In winter, it can take heat from the outside air—even down to quite low temperatures—and pump it into your house. Since it moves heat and doesn't simply create it, a heat pump can actually deliver more Btu of heat to your house than it uses doing the job.

Engineers talk about the efficiency of a heat pump in terms of its coefficient of performance (COP): the ratio of energy that comes out to what goes in. The COP is not a measure of seasonal efficiency but only of operating efficiency under standard testing conditions. The ratio is based on electric resistance heat, which has a COP of one. For every Btu of energy you put in, you get one out. With an electric heat pump, a COP of 2.5 is common; thus you're getting out 150 percent more than you're putting in.

While such numbers sound very good indeed, they don't consider the energy lost when fuel is burned to generate the electricity, or that lost in transmitting it over power lines. Generally speaking, only

## Winter heating

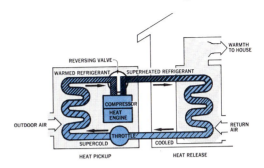

## Summer cooling

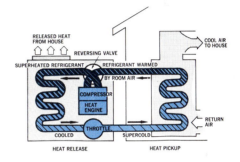

**Gas-fired heat pump** would use energy in gas to drive a compressor and thus "move" heat from outdoors to indoors (top), just as an electric heat pump does. The heat created by the burning gas would also be captured to heat the house. Like an electric heat pump, the gas unit would have a reversing valve to reverse the flow of refrigerant, converting it to an air conditioner in summer. Prototypes of gas-fired heat pumps are now under test. If successful, units could reach the market by the mid-80's.

about 30 percent of the energy content of the fuel reaches your home.

That's why a gas-fired heat pump can offer real fuel savings. Rather than burning the fuel to generate electricity—and wasting two-thirds of it in the process—you would burn the gas on site to power the heat pump.

**Absorption-cycle gas heat pump** produces a heat-pump effect as one fluid (the refrigerant) is sequentially absorbed, boiled out, and condensed in another (the absorbent). A second loop, of water/ethylene glycol, transfers heat to or from the house, depending on season. The heating mode is shown at right. Liquid refrigerant enters evaporator; ambient heat vaporizes it. Vapor enters absorber, is reabsorbed in so-called weak liquid, which becomes a "rich" liquid and gives off heat in process. Rich liquid is

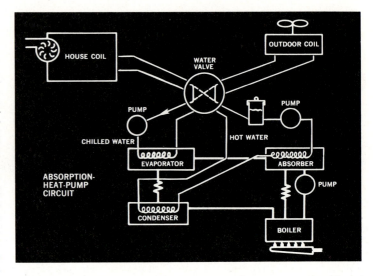

pumped to the boiler, where heat from gas flame separates rich liquid into weak liquid and refrigerant. The refrigerant goes to the condenser and gives off heat. The water loop takes heat from both absorber and condenser and transfers it to the house. GRI/Allied/Phillips scheme is shown.

The Gas Research Institute (GRI) and the Department of Energy have sponsored the development of two residential-sized gas heat pumps. One, a Stirling-Rankine heat pump, is being developed by General Electric. The other is an absorption-cycle device, under development by Allied Chemical and Phillips Engineering of St. Joseph, Mich.

The absorption-cycle heat pump (see diagram) is not a new idea. A major manufacturer tried to bring such a unit to market in the late '60's, but it was plagued with problems. Those have largely been overcome now, says GRI. The Allied-Phillips unit is expected to operate at a COP of 1.3 in the heating mode and 0.5 in cooling, excluding parasitic losses.

General Electric's gas heat pump is quite a different concept. It will use a gas-fired Stirling-cycle engine to power the refrigeration cycle. "The target goals are for a COP of 1.9 in the heating mode," says Dr. James Drewry of GRI, "and 1.0 in cooling." These numbers assume standard testing temperatures and exclude parasitic losses.

Gas heat pumps will be more expensive than other heating and cooling equipment, admits Drewry, "perhaps 20 to 30 percent more than a gas furnace with electric air conditioner. But, with the energy it could save, the payback period would be only about three years," he adds.

If gas heat pumps can be successfully produced, they could save perhaps 30 to 50 percent more energy than even a pulse-combustion furnace. But much R&D and a great deal of testing must still be done before their viability can be determined.

# Oil-burner hazards

*You can avoid sinister carbon buildup*

## By E.F. Lindsley

Squeezing every bit of heat from your oil burner is important these days, but don't forget about safety. Our family learned this lesson last winter when a furnace professional fine-tuned our burner to ex-tract the most heat from every gallon of oil. After his visit, we suffered a winter-long plague of throbbing headaches, accompanied by a strange feeling of listlessness. No one realized until spring that the oil burner was feeding carbon monoxide and other combustion gases directly into the house.

The furnace pro had special instruments to analyze stack gases, draft, and stack temperature. Even so, he made too fine an adjustment of the burner's nozzle position and air supply. The result: poor combustion gases flowing back through the burner. From there, it was a short

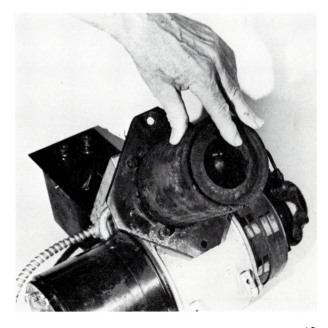

**Newly cleaned** oil-burner gun had to be removed from furnace for scraping after author discovered carbon buildup at nozzle and ignition points. Though parts aren't visible with burner installed in the furnace, running a little stick or wire into the circular gun housing will test for carbon buildup. If huge lumps are dislodged, the burner needs adjusting.

**Oily, fluffy carbon deposits** on slotted air ports of the oil burner's blower mean trouble. A finger check of these ports should reveal only some dust and dirt.

**Smeary black deposits** on inside burner housing and on the blower vanes indicate that combustion gases are coming back into your home through burner.

path to the circulating fan and through the ducts to our living quarters.

You can do a few rough-and-ready checks to see if this hazardous condition is affecting your burner (see photos). Be sure to flip off the main burner switch so it won't kick on while you're checking.

For a more sophisticated check, contact your local air-pollution-control authority and ask to borrow a pollution-detection device. It'll pump up a sample of the questionable air, and measure pollutant content in parts per million. Be sure

to keep the burner running while testing, and take your samples close to the burner air-inlet slots and around the housing. Remember—with the power on, don't put your hands close to the burner to make these tests. The ignition transformer delivers extremely high voltage and deserves respect.

Whichever test you use, if you find carbon monoxide or carbon in the wrong places, call for some skilled oil-burner service.

# Burner adjustments to cut fuel costs

*Now it's possible to spend less on heating than you did last year*

## By Evan Powell

Want to heat your home for less than last year, or maybe even the year before? Despite endless rounds of price escalations for fuel oil and gas, it's possible. You won't do it by bolting on a miracle gadget in a sealed box, but a few modifications and techniques can make for big savings.

Most of these tricks are well known to heating technicians, and the required parts are available from heating suppliers in almost every town. When these adjustments are made as part of normal furnace maintenance and cleaning, cost should be minimal. But you should know what to ask for.

You should also understand something of the basic operation of a furnace. The burner regulates the amount of fuel and air that flows into the furnace and ignites the mixture. The flame is contained within the combustion chamber, which is inside a large compartment called the heat exchanger. The gases flow out of the combustion chamber, through the heat ex-changer, and exit through the flue or vent to the outdoors. While all this is going on *inside*, the furnace blower is circulating air from the house around the *outside* of the heat exchanger. Heat is transferred to room air with no interchange with the flue gases; the heated air flows through the supply ducts and back into the house.

With an oil burner, a pump brings oil from the tank and forces it through a nozzle into the combustion chamber. Both nozzle size and pump pressure determine the amount of fuel that flows into the furnace. Nozzles are classified by the amount of fuel they deliver each hour at a pressure of 100 pounds per square inch.

You probably don't need all the capacity your burner is delivering. Many furnaces were oversized for the space to be heated when they were first installed. If you have since insulated, weather-stripped, and sealed your home, the heat loss is probably far less than the burner rating allows for, so the burner is now even more oversized and less efficient.

"Downfiring" or "derating" simply involves reducing the nozzle size to cut down the amount of fuel the burner consumes. My own furnace was originally equipped with a 1.00 nozzle: At constant operation, the burner would consume one gallon of fuel per hour. I reduced it in stages until now it's equipped with a 0.65 nozzle. This has proved more than ade-

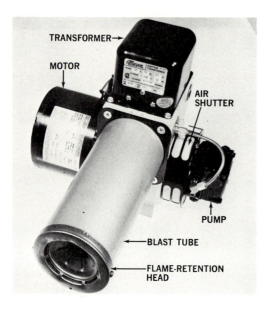

**High-efficiency oil burner** has flame-retention head to ensure more even combustion and afterburning of vapors. Air shutter is adjusted with combustion-test instruments to regulate air-fuel ratio.

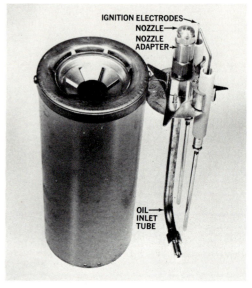

**Concave design** of flame-retention head allows air to swirl into vapor for more even mix. Nozzle assembly, exposed here, fits inside blast tube with nozzle tip behind hole at center of head.

quate. It doesn't mean that I've reduced oil consumption by 35 percent, since the furnace runs longer. But it is more efficient, and the gain is noticeable (from 15 to 20 percent in my case, based on fuel consumed per degree-day).

How low can you go? That depends on several factors. The design temperature of your area is equal to the coldest nights you normally experience during a heating season. At that temperature, a properly sized burner will operate constantly. On the rare occasions when temperature drops lower, the heat level in your home may also drop somewhat. (Another limiting factor is the combustion chamber; more about that in a moment.)

This same technique applies to gas furnaces, except that smaller orifices rather than nozzles are substituted for the present ones, reducing gas input.

With most modern furnaces, you'll find that you can downfire substantially without any combustion-chamber modification. In my previous example, the firing rate is little more than half the original, yet I've had no problem in this area.

For adequate heat transfer, there should be proper flame contact with the surfaces of the combustion chamber, which can be determined by looking at the flame pattern with a special mirror.

If the combustion chamber is too large or if it's damaged, special "wet pack" repair kits are available. Using this kit, the ceramic material is molded onto the inner surface of the chamber, then the burner is fired to harden it. The resulting ceramic fiber material is durable and efficient.

The nozzle regulates the amount of fuel passing into the burner, but air is also important to the final mixture. The air flows into the burner through a passageway called the blast tube. For top efficiency, the fuel/air mixture must be consistent and must occupy the proper volume of space. Modern turbulators, or flame-retention heads, introduce the air to the fuel vapor in a swirling pattern that helps promote even burning. Older burners can often be fitted with a new head to improve their efficiency.

At the same time these modifications are made, the inner heat-exchanger surfaces, smoke pipe, and chimney should be thoroughly cleaned. Soot can insulate and, in severe cases, restrict the flow of the flue gases through the heat exchanger, thus reducing the heat transfer to the room air.

The result of all these modifications can be evaluated with combustion-test instruments. Tests should include a check of flue temperature, draft, smoke density, and $CO_2$ level of the flue gases. At each step, the technician should adjust the burner to the highest possible level of efficiency.

The next time you call for furnace maintenance, tell the servicing agency what you expect and ask about its experience with upgrading the efficiency of burners. If your furnace is more than five years old and you have reduced your home's heat loss, these modifications should help you beat last year's fuel bill.

# Automatic flue dampers

*How safe are they?*

## By Evan Powell

Since the introduction of automatic flue dampers, a lot has happened—not so much to the product as to the regulations that govern their installation and to industry-wide safety standards. During this time, also, claims made for many of the devices have become less exaggerated.

Dampers help reduce some of the inefficiency of the furnace. During normal burning operations, about 25 percent of the heat produced goes up the chimney with the flue gases. But heat loss doesn't stop there. When the furnace shuts down, the thermosiphoning action of the heated air rising up the chimney can continue, even with air at room temperature.

That's where the add-on flue damper comes in. Installed in the smoke pipe, a damper blade automatically closes to stop the flow of air up the chimney when the furnace is not operating. A switch or relay is incorporated in the damper mechanism to prevent operation of the burner motor or gas valve until this damper reopens.

The units I've checked out are reasonably well constructed and incorporate the proper safety devices. They are also listed as having met the safety standard of Underwriters Laboratories—a standard that was nonexistent at the time of our previous tests. The American National Standards Institute and the American Gas Association have also established standards for vent dampers and their installation. The dampers' track record for safety has been good so far.

Though the vent dampers themselves seem to be safer than when they were introduced in 1976, it's still possible for

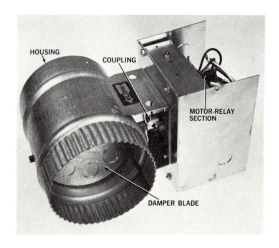

**Flue damper**—a simple device with blade inside steel housing—is designed to replace short section of smoke pipe. Relay prevents operation of oil-burner motor or gas valve until damper is fully open.

**Dampers don't seal** opening in smoke pipe entirely, in order to allow for some burning after shutdown. This gas model has hole in center for pilot exhaust. Additional knockouts are provided in case smoking occurs in test operation.

them to be improperly installed. They're not marketed or intended for do-it-yourself installation. But even when a pro does any flue-gas job, check his work.

The actual saving you can realize depends on the present installation and location of your furnace. The damper will do nothing for you during the normal operating cycle of the furnace. However, many furnaces are off substantially longer than they're running, especially in mild weather, and the damper can save heat loss *if* the furnace draws upon room air.

There's an alternative to vent dampers. If outside combustion air is available to the furnace burner, no scavenging for conditioned room air will occur during the off cycle, although the flue damper could still have some slight advantage by reducing the cool-down of the heat exchanger. And downfiring the burner will also net you some savings when the burner is in operation.

In most areas, a vent damper will cost from $250 to $500, installed. I'd recommend installing a double fail-safe device at the same time—an oil-inlet-line solenoid valve for an oil furnace or an additional gas valve for a gas furnace, to insure that the burner can't come on with the damper closed.

# Heat pumps

*Do they make more sense now for more homes?*

## By V. Elaine Smay

If you still think of the heat pump as an offbeat, backward air conditioner suitable only where winters are mild and short, it's time to take a fresh look. The heat pump's popularity is growing rapidly, even north of the Mason-Dixon line. About 50 manufacturers now make heat pumps and some one million units are sold annually.

A heat pump is basically an air conditioner with a reversing valve that switches the flow of refrigerant so that it can provide heat as well as cooling (see box, page 52). It uses electrical energy to move heat from a colder to a warmer location. But moving heat with electrical energy is more efficient than converting electrical energy into heat, as do other electrical heating systems. The heat pump's comparative efficiency, the changing energy picture, and improvements in the hardware are the reasons for its growing popularity, and why you may want to consider one.

## THE ENERGY PICTURE

Gas and oil have long been the most popular fuels for home heating. But oil is often scarce in some areas and expensive everywhere, and new gas hookups are not available in many places. Thus, with a new house you may have only one resort: electric heat. Often, this means electric-resistance heat—the kind that, like a giant toaster, heats your house with hot wires. Since electricity can be generated from coal, nuclear energy, or hydro-electric power, its future availability is more certain than that of gas or oil. But electric-resistance heat is expensive and getting more so as electric rates rise.

## HEAT-PUMP ECONOMICS

Engineers speak of the efficiency of a heating device in terms of its coefficient of performance (COP). The COP is the ratio of energy, put in (kWh are converted to Btu per hour) to the heat energy put out (in Btuh). With resistance heat the COP is always one: You put in one Btu, you get out one Btu. But since a heat pump moves heat rather than just converting electrical energy to heat, you get more Btu delivered to the house than the Btu equivalent of the electrical energy used to deliver it. The COP figures manufacturers give are de-

**Split-system heat pumps** have an outdoor and an indoor unit. Carrier's Weathermaster III has separate compressor inside the building to assure reliability in cold climates, says company. Compressor is usually in outdoor unit.

rived from standard testing procedures, which measure the COP at 70° F inside temperature, and 47° and 17° outside temperature. A first-rate residential heat pump might have a rated COP of 2.8 at 47° and 1.9 at 17°. Most are somewhat below that.

COP is higher with warm outdoor temperatures because it's easier for the heat pump to extract heat from, say, 50° F air than from air at 40°. Once the outside temperature drops below about 40°, another factor comes into play. The outdoor coil tends to chill to below the freezing point and accumulate frost. High humidity and strong winds accelerate frost formation. So periodically the heat pump reverses itself and becomes an air conditioner for a couple of minutes. This circulates warm refrigerant to the outdoor coil and melts the frost. As the machine

switches from heating to cooling and back again, there are cyclic energy losses, which decrease its efficiency.

That's not all: Virtually all heat pumps have backup heat, usually electric resistance. When the heat pump goes into a defrost cycle the backup heat comes on. That chips away a bit more at your operating efficiency. Still, the heat pump performs very well down to about 30° and can supply all the heat your house needs, except during those brief defrost cycles. But as the temperature of the outdoor air drops lower and lower, the heat pump can extract less and less heat from it. At the same time the heat loss from your house goes up. At some temperature, called the balance point (see graph, page 52), the heat pump can no longer supply all the heat. Below the balance point (25° to 35° in most houses) the backup heat is cycled

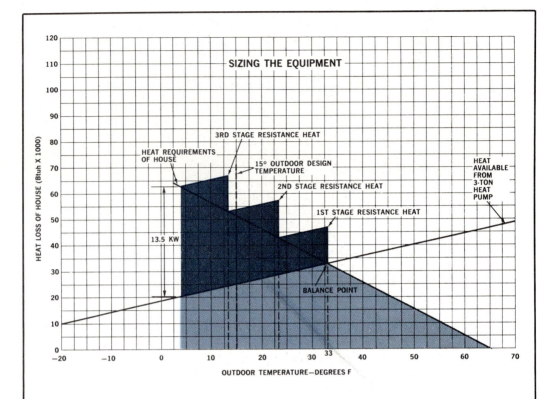

**The size heat pump to be used** is generally based on a building's cooling needs. It may be oversized by 25 to 35 percent in colder climates to take care of more of the heat load, but if the pump is too large, it will cycle on and off frequently in summer and do a poor job of dehumidification. In this example, the contractor selected a three-ton heat pump. Next he sizes the backup heat: on a graph he draws a line representing the heat requirements of the house (based on his detailed energy analysis) and another for the heat available from the heat pump. The point where the two lines cross is called the balance point—usually between 25° and 35° F. (Some electric utilities mandate a minimum balance point, which often requires a larger pump.) Above the balance point, the heat pump can supply the total heating needs of the house. Below it, backup heat, usually electric resistance, supplements that heat pump. The backup heat will usually be actuated in two or more stages, each controlled by an outdoor thermostat (see text). The house here will have 13.5 kWh of back-up heat—80 percent of its total heating needs. In some places the utility may require 100-percent backup.

on and off to supplement the heat pump. The colder it gets outside the more resistance heat you have to buy.

Many heat-pump makers calculate what they call the Seasonal Performance Factor (SPF), as well as the COP. The SPF attempts to consider the operating efficiency of a heat pump over a wide range of outside temperatures and relate that to local climates. Manufacturers generally claim that the SPF of a good heat pump ranges from about 1.5 to 2.5 in the continental United States. But both COP and SPF are based on laboratory tests.

Tests that measure heat pumps in people's homes usually indicate somewhat lower operating efficiency. But whatever the actual numbers, heat pumps look very good when compared to resistance heating. A fact sheet prepared by Oak Ridge National Laboratory for the Department of Energy (DOE) gives these conclusions: "While actual savings depends upon such factors as climate and the price of . . . electricity, heat pumps offer an average of 20-percent savings over conventional cooling systems and central electric-resistance heating . . . . In some regions of the country, a heat pump can reduce electric bills by 35 to 45 percent."

## LIFE-CYCLE COSTS

But a heat pump costs considerably more to buy and install than electric-resistance heat, so you need to look at life-cycle costs to determine if there's real economy.

Using a computerized energy analysis, a Carrier dealer in the Cleveland area might figure that a heat pump for a new 1,800-sq.-ft. house would cost $3,200 installed. An electric-resistance furnace with central air conditioning would cost $2,700. The annual savings with the heat pump, based on the current cost of electricity for heating in Cleveland and average degree-day data, would be $417 ($741 for the heat pump vs. $1,158 with resistance heat). Monthly savings would be $34.75, assuming cooling costs the same with both systems (some sources claim that a heat pump will be about 10 percent less efficient than a standard A/C; others deny that). If the buyer had a 25-year mortgage at 9.25 percent, a heat pump, costing $500 more, would add $4.28 to the monthly mortgage cost, according to the

Carrier computer. Thus, the real monthly savings would be $30.47. The heat pump would then make up the difference in purchase price in less than 1½ years.

These figures are based on normal winter weather, when a heat pump in the Cleveland area would be expected to use 30 to 40 percent less electricity than resistance heat. But during a recent winter, which was extremely cold, the Kopf Construction Company measured actual electricity usage in 20 of its houses in the suburb of Avon Lake (10 had heat pumps; 10 had baseboard resistance heat) and found that from November through April the real savings with the heat pumps averaged only 16 percent per month. So payback time isn't always predictable.

Without air conditioning, the payback period would be much longer. Baseboard resistance heat (cheapest to install) might

**Environmental chamber** at Carrier testing lab can simulate about any weather. Some use snow-making equipment, like a ski resort. Various sensors on heat pumps monitor critical components and feed information to computer.

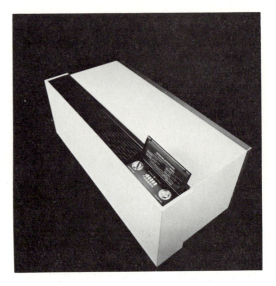

**Through-the-wall heat pump** from General Electric, called Zoneline III, conditions single room. It could heat and cool a new addition or vacation cottage. There are three models: 8800, 11,400, and 14,200 Btuh (at 47° F). Rated COP's: 2.1, 2.0, 2.0 (at 47°).

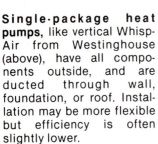

**Single-package heat pumps,** like vertical Whisp-Air from Westinghouse (above), have all components outside, and are ducted through wall, foundation, or roof. Installation may be more flexible but efficiency is often slightly lower.

**Outdoor unit** (right) looks like that of regular split-system heat pump. Add-on heat pumps cost a bit more than central A/C, but the added expense may be quickly recovered through savings in heating costs.

**Factory training** helps insure that heat pump is installed and serviced by qualified contractor. Major heat-pump makers all train and certify dealers. BDP Co. trains Bryant (shown), Day & Night, and Payne dealers (all its trademarks).

## IMPROVED PUMPS

Today's heat pump is much more than a reversible air conditioner. The good ones incorporate important design modifications that have made them much more reliable than their predecessors. "Most of the major brands are extremely reliable now," says Barney Menditch. "They still may have little problems, but that can happen to anything, even a gas furnace."

The efficiency as well as the reliability of the heat pump has increased. It's now a better heating machine—the COP's are up. It's also a better cooling machine. In the cooling mode, efficiency is stated in terms of the Energy Efficiency Ratio (EER), as with an air conditioner. The higher the EER, the more efficient the heat pump. An EER of seven is pretty good, eight is very good. Many high-efficiency heat pumps today have an EER of nine or above, better than most air conditioners and only slightly below the best A/C.

Today you can buy a high-efficiency heat pump or an economy model. The difference in price will likely be $200 to $300 (installed), says Barney Menditch. A heating contractor can tell you the payback, based on the reduced operating costs of the high-efficiency model.

cost $900 for this hypothetical 1,800-square-foot house in Cleveland. A heat pump (at $3,200) would then have a payback period of 10 to 15 years—assuming normal weather.

A heat pump can even compare favorably with an oil furnace with central air conditioning. Barney Menditch, president of General Heating Engineering Company, near Washington, D.C., says, "In this area the heat pump is five to 10 percent cheaper than an oil furnace and air conditioner, both to install and operate."

A gas furnace and central air conditioning, however, generally cost less to install than a heat pump: "From a standoff to 20 percent less," says Edwin Douglass, manager of space conditioning and applications for the Edison Electric Institute in New York.

The cost of replacing a standard heating-and-cooling plant with a heat pump could run as high as $3,500, according to the Oak Ridge/DOE study, if your wiring and ductwork have to be beefed up.

## NEW EQUIPMENT

A few manufacturers now make add-on heat pumps, designed to work with an existing forced-air furnace: electric, oil, or gas. An add-on heat pump is especially attractive if your present furnace is healthy and you want to install (or replace) central air conditioning. Depending upon exact conditions, an add-on heat pump, which may cost $300 more than central air conditioning, can pay back that cost quickly, especially if it's used

## How a heat pump works

A heat pump is a refrigeration device, like the fridge in your kitchen and the air conditioner in your home and car. All refrigeration systems make use of the same physical principles: 1) As a liquid evaporates and turns to a gas, it absorbs heat. 2) Conversely, as a gas condenses and becomes a liquid, it gives up heat. 3) Whether a fluid is liquid or gaseous depends on temperature *and pressure*.

In a heat pump (or any refrigeration system), a refrigerant (a fluid that boils at a low temperature) is subjected to varying pressures that change it from a gas to a liquid and back to a gas at given operating temperatures. Consequently, it can gather heat from one location and give it up at another.

In the cooling mode, the refrigerant evaporates in a set of coils located some- where in the duct system of your house, thus absorbing heat (like an ordinary air conditioner). This gaseous refrigerant then flows to coils outside your house where it condenses, giving up that heat. Unlike an air conditioner, however, a heat pump has a reversing valve that can make the refrigerant flow in the other direction. That turns it into a heating machine. In the heating mode, the refrigerant evaporates in the outside coils, gathering heat from the outdoor air (which contains some heat all the way down to absolute zero: −460°F.) Then the refrigerant condenses in the coils inside your house, releasing the heat. Whether it's heating or cooling it moves heat from a colder to a warmer location. It uses electrical energy to do this.

**New add-on heat pumps** don't replace existing electric or fossil-fuel furnace; they use it for backup heat. Indoor coil of York's Maxi-Mizer (above) is installed over furnace, like central air conditioning, and will usually use existing air handler and ductwork.

with an electric furnace. It may also be cost-effective on an oil furnace. And used with oil backup, a heat pump can prove economical in very cold climates where one with resistance heat might not.

The actual installed cost of an add-on heat pump depends on whether you have to put in a new electric service entrance (a heat pump requires at least a 100-amp circuit), whether your present air handler would have to be replaced, and whether your ductwork is good enough. A heat pump delivers air at a lower temperature and lower velocity than other central-heating plants. Consequently, it has to deliver more of it.

"You probably could install a 2½-ton heat pump over an electric furnace for around $1,600 in the Washington area," says Menditch. "If you had to put in a new air handler it would run maybe $1,900." He didn't even want to talk about new or modified ductwork. The possibilities are too variable.

# A HEAT PUMP FOR YOUR HOUSE?

Since a heat pump is both a heating and cooling machine, its higher first cost probably can't be justified in a really warm climate where little heat is needed. Likewise, where you don't need—or want—cooling, the heat pump is probably not for you. But in most of the country—from northern Florida to as far north as Chicago—heat pumps may well pay off.

Only a qualified heating contractor can accurately evaluate your house, assess your local climate, plug in the costs of available energy, and tell you whether a heat pump would be a good investment.

Heat pumps today that are made by the major manufacturers are highly reliable machines. But as John Callor of Carrier puts it: "There are still some dogs on the market." Here are some things you should know to help you avoid the lemons and get the most out of your purchase.

- *Warranty.* A standard factory warranty (from a good manufacturer) will cover all parts for the first year. The contractor who installs the system will furnish labor. The factory warranty covers the compressor for five years. The homeowner pays for other parts and for all labor from the second year.

- *Service contract.* Manufacturers of the best heat pumps offer a national service contract which is essentially an insurance policy for the second through fifth year. If you buy the service contract (for something like $50 per year), all parts and labor are covered for this time period. The contractor may also offer semiannual inspections for an additional fee.

The Alabama Power Co. has collected years of data on heat-pump reliability (which is not made public for legal reasons), and advises: "Buy from a manufacturer that offers a national service contract. All the good ones do."

- *Features.* A heat pump should have a filter/drier in the refrigeration line to filter out moisture and dirt. It should also have a crankcase heater. This boils off liquid refrigerant in the compressor, where only gaseous refrigerant should be.

- *Options.* Outdoor thermostats are a good investment, except in warm climates. These insure that the backup heat does not come on unless the temperature outside drops below a certain level. Another good option is a warning light on the indoor thermostat that alerts you if you're using backup heat when the system is calling for the heat pump. Without the warning light, you might not know you were on backup—until the bill came.

- *The contractor.* Selecting a competent contractor is also very important. Edwin Douglass of the Edison Electric Institute offers these suggestions: Choose a contractor who has had factory training. Find out how long he has been installing heat pumps and ask for a list of satisfied customers. Electric utilities can also guide you.

# Ground-water heat pumps
## Home heating and cooling from your own well

*If you're sitting on "good" water you may cut your fuel bills dramatically*

## By Robert Gannon

If you're building a home or just planning to replace your furnace, a ground-water heat pump may be what you're looking for.

A what?

A ground-water heat pump is a device that cools water—usually from a well—and then pumps the extracted heat into your home. In summer, it withdraws heat from the inside air and uses water to carry it away.

Do you have a large supply of good water, live in a reasonable climate, and plan to install central air conditioning as well as a new furnace? If so, you could use a ground-water heat pump—and chop a third, a half, and maybe even more from your heating and cooling bill.

The idea works. Good equipment is available. It ordinarily requires little maintenance, costs not much more than conventional furnaces, and amortizes itself in only a few years.

So why aren't more people using the system? The question has a number of answers, and they're confusing. Ask someone in the trade about ground-water heat pumps and you're likely to get answers that are uninformed, misinformed, or downright wrong.

To try to arrive at the truth, I've spent the last few months talking with researchers, manufacturers, contractors, engineers, and homeowners.

Finally, I believe I've sorted out the main factors, and I've arrived at a conclusion: If you can use one, go ahead. You'll probably save yourself a heap of money. But that *if* is a muddy one.

First, some background: Most homeowners know at least something about conventional heat pumps, the air-to-air or air-source models. Essentially, they're air conditioners that can reverse. In summer they cool the house; in winter, they warm it by extracting heat from the outside air.

And the beauty is that when the outside temperature is moderate, the only cost is for the electricity to run the heat pump's compressors and fans—there's no burning of fuel. Such systems can be extremely efficient. Heating units are rated by their coefficient of performance (COP). This is based on electrical-resistance heating in which one kilowatt provides 3,412 Btu in an hour, for a COP of one. A

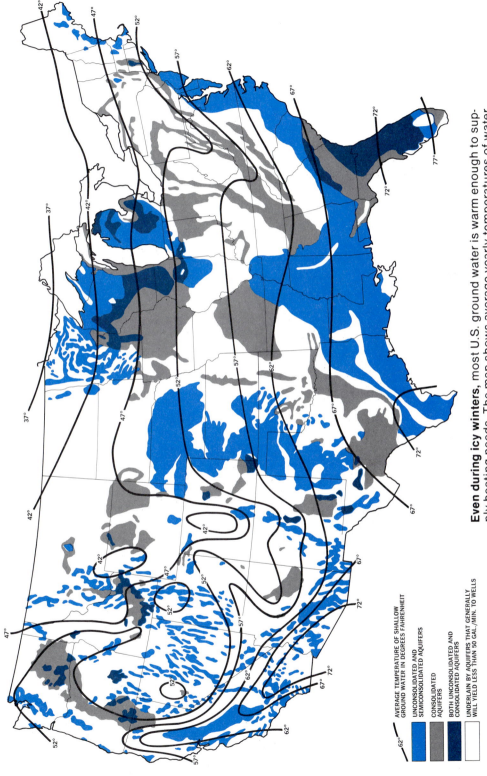

**Even during icy winters,** most U.S. ground water is warm enough to supply heating needs. The map shows average yearly temperatures of water in major aquifers. Every aquifer will yield some water, but there's more flow from unconsolidated aquifers where water is stored in porous layers of sand, gravel, and clay. In semiconsolidated aquifers, silt, "cemented" by minerals, partially blocks the water's flow. Cementing process is more advanced in consolidated aquifers.

AVERAGE TEMPERATURE OF SHALLOW
GROUND WATER IN DEGREES FAHRENHEIT

UNCONSOLIDATED AND
SEMICONSOLIDATED AQUIFERS

CONSOLIDATED
AQUIFERS

BOTH UNCONSOLIDATED AND
CONSOLIDATED AQUIFERS

UNDERLAIN BY AQUIFERS THAT GENERALLY
WILL YIELD LESS THAN 50 GAL./MIN. TO WELLS

59

heat pump not only produces heat, but moves it from one place to another, so the COP of an air-source system can be much higher than one—even three, when the outside temperature is around 50°F. In other words, for each watt of electricity consumed by the unit, three watts of heat energy are available for warming the house—one from the unit, two from the outside. And the electric heating bill drops by two-thirds.

But the efficiency of air-source systems plummets as the outside temperature falls. As temperatures approach freezing, such severe strains are put on the units that they automatically switch in resistance heat. A similar problem arises in summer when temperatures edge into the 90's. As the amount of electrical energy needed to move heat outside mounts, the efficiency of air-source units dives.

An ideal situation, of course, would be one in which outside air remains at a constant temperature.

And that's where ground water comes in. Unlike air or surface water, ground water (from a well or spring) has a stable temperature: the mean annual temperature of the overlying air (see map). In winter, ground water is always warmer than the air; in summer, always colder.

So you drill a well and tap this water, using it as a heat source in the winter, a heat sink in summer. Simple. The end result is a higher COP which remains constant no matter how far outdoor temperatures plummet—so constant, in fact, that in most installations no backup heat is required.

The idea isn't new. The concept has been kicking around since not long after World War II. But hardly anybody outside the Deep South has been interested—for largely the same reason that you didn't hear much about solar collectors before the oil embargo. Conventional home heating was cheap. Why experiment?

## GROUND-WATER PIONEERS

One who did experiment is physicist Carl Nielsen, a professor at Ohio State University. He has one of the oldest continuously operating water-source heat pumps on record.

"I learned the principle of heat pumps when I was a physics student," he told me. "So when we first built, in 1948, I decided to try a water-source pump. It's clean, I thought; it doesn't produce soot; I don't have to put up a chimney. And it's relatively quiet; I didn't want an oil furnace that would roar at me."

No commercial ground-water heat pumps were available then, so Nielsen devised his own one-ton unit, using standard refrigeration-plant parts, except for the heat exchanger. "For that, I took two lengths of pipe—a one-inch-o.d. piece and a ⅝-inch-o.d. piece—pushed one inside the other, then coiled them up. Worked fine."

He drilled an 80-foot well in the backyard, determined that the 44-degree water had a sufficient flow, and ran his outflow to a pond alongside the house. The outflow, by slow percolation, would eventually return water to the well.

The unit ran for seven years with no trouble. So when he built a larger house on the same property, he again planned for ground-water heat. This time he built a two-kW unit. It's been running ever since. And the problems he's had over the years can be narrowed to one: Several years ago a starter relay burned out.

One severe problem, I had heard, is scaling; another is encrustation. In the beginning, Nielsen was concerned, too. His well water is extremely hard and rich in iron. "If you fill a jar and let it stand for a day, it turns brown," he said. "I wondered what it would do to my coils." Nielsen unbolted the end of his heat exchanger. "Here, stick your finger in." I

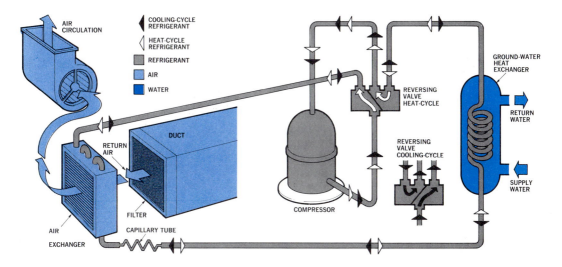

**Like an air conditioner,** a ground-water heat pump uses circulating refrigerant to absorb heat. Unlike an A/C, heat is absorbed from water, not air, and is released inside the house. In a typical heating cycle, liquid refrigerant flows through a capillary tube that lowers its pressure and boiling point. Passing through the ground-water heat exchanger, the refrigerant extracts heat from the circulating water, boils and vaporizes. The cooled water returns to the ground while the warm, low-pressure gas travels to the compressor. There it's squeezed to form a high-pressure, hot gas. Pumped through a reversing valve to the air heat exchanger, it condenses, releasing heat to the circulating air. The warm air is ducted throughout the house, and the liquid refrigerant flows back through the capillary tube to repeat the cycle. For cooling, the process reverses. The compressor sends the hot gas directly to the water heat exchanger to release heat collected from the house. The warmed water returns to the ground while the cooled, liquefied refrigerant flows through the capillary tube to the air heat exchanger. There, it again absorbs heat from the house and vaporizes. The gas returns to the compressor, which pumps it back to the water heat exchanger to renew the cycle.

did—and got only a light brown smudge of oxidation. "The coil was built so that any buildup of scale could be cleaned out," he said, "and I bought a wire brush to be used with my hand drill. But here it is, unused."

When Nielsen installed his pump, oil in Columbus was 15 cents a gallon, electricity two cents a kWh. Now the prices of both have zoomed. "With the 3.5 COP I'm getting, the cross-over came when oil hit about 25 cents," he said. "At that point the pump was no longer simply an experiment; it was a money saver."

**Dr. Carl Nielsen** shows off the ground-water heat pump he designed and built in 1955. Today, it still provides efficient heating and cooling for his 2,000 square foot home. Behind him, heat-exchanger pipes circulate well water and refrigerant. Constant-temperature water acts as winter heat source, summer heat sink.

## PUMPING INTO A THINK TANK

While heat-pump engineers over the years have largely ignored single-family residences, they have developed effective ground-water systems for large buildings. Not far from the Nielsen home, in downtown Columbus, is Battelle Institute, a scientific think tank. A 317,000-square-foot section of the sprawling complex is heated and cooled entirely by heat pumps—probably the largest setup of its kind in the country. Battelle gets its water, at 54°F, from five 16-inch wells drilled to 50 feet in a sand-and-gravel aquifer, and each month pumps some 40 million gallons through the exchanger and out into nearby Olentangy River, which supplies the aquifer.

The massive equipment works pretty much the same as the unit in Dr. Nielsen's basement, but the physical difference is striking. Nielsen's setup is stuck off in a dingy recess and takes up about as much space as a freezer. Battelle's controls alone occupy a complete room—its walls are decorated with huge flow charts and lined with digital counters flicking remote readings.

Battelle developed the system in the days of cheap energy, simply because its engineers were intrigued with the idea. Today, it's one of the best examples around of how efficient ground-water heat pumps can be: The heating COP in one of the buildings is 5.4.

In the years since Battelle installed its systems, ground-water heat pumps have become commonplace in large buildings: They now number in the thousands.

But home units haven't kept pace. Now, though, some experts see a dramatic growth on the horizon. Says Charley Smith, a spokesman for York Division of Borg-Warner: "Quite frankly, until the energy situation turned around, sales of water-source units were marginal, but they're really starting to pick up now." And at least two manufacturers, WeatherKing and Vaughn, predict that sales will double.

Just how much can today's homeowners hope to save with a ground-water heat pump? In a 1976 study prepared under a grant from the U.S. Environmental Protection Agency, the National Water Well Association stated that "at present prices, a homeowner can have his investment for a [water source] heat pump returned from cost savings resulting from reduced energy consumption in four to eight years [compared to] conventional heating and cooling units" (see box, page 64).

With oil prices surging and coal largely unacceptable to most Americans, groundwater heat pumps seem to make sense. But if so, why don't many more homeowners have the units? The objections, I've found, boil down to five.

## PROBLEMS: REAL AND IMAGINARY

### You need a well

"Sure, it's an effective device and an energy conserver *if* you've got the water," says Ben Sienkiewicz of the Air Conditioning and Refrigeration Institute (ACRI). "But how many of us have a well in our yard?"

A few families are using lakes, rivers, or canals, but for most homeowners, a well is necessary. And that means an average of about $2,000 in drilling costs, according to the National Water Well Association.

You'll need a minimum flow of about 2½ to three gpm per 12,000 Btu needed. Your chances of getting that are good.

Hydrologist Jay Lehr, executive director of the National Water Well Association, says that if you randomly sink holes to two hundred feet, about 80 percent of the time you'll find a flow rate of three gpm or more. But you may need another well to return the water to the aquifer.

Heat-pump users with an unlimited supply of water often dump the outflow into a creek or pond, or even into the nearest sewer. The movement, though, is toward recharging: returning the water—slightly warmed or cooled—to the ground

via a second well. In Bexley, Ohio, two next-door neighbors each drilled a well and now share them, drawing from one and discharging into the other. If the water ever begins to run low, they'll simply reverse the flow.

Actually, the discharge well needn't be a conventional drilled well, complete with casing. A bored hole that's filled with coarse gravel to keep the sides from caving in and that's two or three feet in diameter and 20 or 30 feet deep (depending on the percolation rate) costs only about a fifth as much as a drilled well. And some people use the outflow to fill ponds or simply to water grass—a form of recharging.

### Ground water is mysterious

"There's a built-in prejudice against ground water simply because you can't see it," says hydrologist Lehr. "The reason air-source heat pumps are so widespread is not that they're better. They're not. It's simply that air is right there, surrounding us. You're not really sure the unseen ground water is there."

As evidence of this prejudice, Lehr points out that Americans use three times as much surface as ground water, even though the cost of developing groundwater supplies is only one-tenth that of reservoirs.

### High initial cost

A heat pump, either water- or air-source, costs at least 50 percent more than a conventional furnace. Add to this another $1,500 to $2,500 for a well and you've doubled the initial investment. In addition, your ductwork will have to be larger if you use forced-air heat. A furnace delivers air that's 160°F. A heat pump delivers air at only about 120°, so it must move a larger volume. You need about twice the normal-size ducts (or double the number) to accommodate the flow, with a larger

| Heating System | | Initial cost | Annual amortization | Annual fuel cost | Annual heating cost (nearest $25) |
|---|---|---|---|---|---|
| Gas furnace | | $ 700 | $ 85 | $308 | $ 400 |
| Oil furnace | | 1400 | 170 | 475 | 650 |
| Coal furnace | | 1400 | 170 | 425 | 600 |
| Electric furnace | | 900 | 110 | 880 | 1000 |
| Air-source heat pump | (a) | 3000 | 456 | 463 | 925 |
| w/air cond. | (b) | 800 | 122 | 463 | 575 |
| Ground-water heat pump | (a) | 3000 | 456 | 275 | 725 |
| w/air cond. | (b) | 800 | 122 | 275 | 400 |

Physicist Nielsen worked out the table above using figures gathered in Columbus, Ohio, during the winter of 1977. The figures are probably typical of costs throughout the central belt of the U.S. and, of course, are much higher today.

The table assumes an annual heating requirement of 100 million Btu. Nielsen assumed an annual average COP of 3.2 for the ground-water heat pump and 1.9 for the air-source heat pump, and estimated high for the initial cost of this equipment. For fuel costs, he figured gas at $2 per 1,000 cu. ft., oil at 42¢ a gal., coal at $80 a ton, and electricity at 3¢ per kWh. Fuel efficiency is rated at 65 percent. Amortization was computed at nine percent interest over a lifetime of 15 years for the furnaces, and 10 years for the heat pumps (though Nielsen says 10 years is probably too short). Maintenance, repairs, and well-drilling costs aren't included. Among the variables are equipment condition, design, manufacturer and installation, plus weather and ground-water temperatures—so the figures are approximations only.

At first glance, it might seem as if ground-water heat pumps offer clear savings only over electrical-resistance heating. But, comparing line (b) with line (a) makes the picture clearer. Line (a) is the homeowner who doesn't install central air conditioning. Line (b) is the homeowner who installs an air-conditioning system as a matter of course (common practice in the South and West) no matter what type of heating unit is used. In this case, the $800 initial cost is the price of adding heating capability to a central air-conditioning unit (i.e., a central air conditioner would cost about $2,200—for $800 more you can get a ground-water heat pump for both heating and cooling). Since $800 is lower than you'd pay for most furnaces, and since annual operating costs are so much less than those for conventional systems, the savings are substantial.

fan turning at fewer rpm. A faster fan with the same ducts won't do; the efficiency would drop, the noise increase.

Similarly, hot-water baseboard tubing must be able to handle about twice the volume, or have twice the tube surface exposed to the air.

# STATE OF THE INDUSTRY

## Nobody's pushing the idea

"Ninety-five percent of the population has never heard the term 'heat pump,' " says Paul Sturges, a Stone Ridge, New

York, heating-cooling consultant. "And the concept of a ground-water heat pump is as foreign as, oh, heating with cow dung or antimatter."

In fact, few heating contractors have studied ground-water heat pumps. They're unfamiliar with the fundamentals of ground water and vaguely uneasy working with something 50 feet under the surface.

Even the manufacturers aren't pushing the idea. Makers are either relatively small and regional, or they produce a whole line of heating equipment, including the big money-making air-source units. Says ACRI's Sienkiewicz: "The volume of water-source pumps is so damn small, and the rest of the industry is in the hundreds of millions of dollars—you devote your time and effort to the high-volume business."

## Some water is unusable

There are two problems here: temperature and quality. The colder the water the more work the pump must do in the heating cycle. Cool water cuts the COP and requires a larger pump and compressor.

Until recently, most manufacturers limited sales to areas where ground water is at least 60°F—and that eliminates most of the upper two-thirds of the country.

WeatherKing general manager Jim Brownell says that his company is gradually extending its range "while we gain experience in what low water temperatures the units can really stand." He recommends that no unit be installed in water below 55°F without a protective thermostat to turn it off if the ground water approaches freezing.

Physicist Nielsen blames the manufacturers, calling the cold-water question a "pseudo problem." He adds: "The notion that it's difficult to design for low temperatures is just so much baloney—and a fail-safe system that turns itself off if

water pressure falls is very simple to include. I did it myself."

Vaughn Company, for one, has designed its standard models to operate with water as low as 40°F, while special units are available to handle water near freezing. Some equipment, in fact, pumps water from frozen-over lakes. But because of heavy-duty components and other extras, Vaughn models, installed, cost about a third more than those of most other manufacturers.

The other water problem, and by far the most controversial, is quality. Iron, calcium, and magnesium salts, hydrates, suspended solids—all can result in corrosion and scale or encrustration.

What are your chances of finding unsuitable water? Robert Ross of the Better Heating-Cooling Council estimates that unusable water will be found under 15 or 20 percent of the land in the U.S.

If your property happens to fall into that group, you may be headed for disaster. One Austin, Texas, contractor will never try a water-source heat pump again: "I've had nothing but bad experiences with them; I put in three units and had so much trouble with scaling that I eventually decided to abandon them."

And in Sarasota, Florida, some homeowners must flush the refrigeration coil with acid every year or two to wash out scale, and eventually the acid itself eats through the tubing. (Strangely, almost no residential units have coils designed the way physicist Nielsen's are—so that they can be cleaned with a wire brush.)

One way to prevent scale is to use cupronickel instead of copper tubing. Some manufacturers sell it for an extra charge; A few provide it as standard. Essentially what happens, says Vaughn's Pirrello, is that the cupronickel expands and contracts with temperature, and its surface tends to flake off mineral deposits and scale with each cycle.

Design could have a lot to do with scaling, too. At least that's what Pirrello claims. The main problem, he says, is caused by internal buffles and fins—which the Vaughn unit doesn't have. What percentage of water does he find unusable? "Essentially, none."

If you decide to look into the possibilities for your home, how can you make a decision? First, you should get some idea of what a well in your yard might produce in the way of volume and quality. One way to determine potential is to ask a large, local well driller for an estimate—usually a free service. (The National Water Well Association will give names of knowledgeable drillers in your area. Their address is 500 West Wilson Bridge Road, Worthington, Ohio 43085.

Local water-softening people might also have some thoughts. A plumber who deals in water heaters could give you an idea of how much scale he's finding. Engineer Owen suggests that if your neighbor has a well, get the water tested at Sears or by the local public health office. Ideally, he says, the pH should be between six and eight, and the hardness reading (on the standard scale of 0-30) no higher than 10.

Next step: Call in a heating-cooling contractor, preferably one with considerable ground-water heat-pump experience. Ask him for the names of some of his heat-pump customers and see how satisfied they are.

The results of your research should help you decide if a ground-water heat pump is for you. Conflicting claims, lack of thorough scientific studies, and the fact that technical expertise is limited to only a few people, make its installation, in my opinion, still something of a gamble. Nevertheless, because I know that my 55-degree water is fairly soft, when my oil-guzzling furnace begins to fail, I plan to take that gamble. Maybe even sooner.

# Turbine-drive heat pump doubles heating/cooling efficiency

*Its heart is a tiny, virtually maintenance-free turbine*

## By David Scott

David Strong pushed the button on a home-heating rig that could cut fuel consumption and bills in half. A miniature turbine wound up to an incredible 150,000 rpm—or so he told me. I couldn't hear it. "You'd need a stethoscope," he said, grinning.

This tiny, silent turbine is the heart of a revolutionary domestic boiler that combines a heat pump with a vapor generator to get super-efficient use of fossil fuel.

A heat pump normally extracts and upgrades free heat from ambient air or water, but needs mechanical energy—usually from an electric motor—to do it. But since that energy is produced with a fuel-to-heat-energy conversion efficiency of no more than 34 percent, the usual heat pump is not much better than a conventional oil- or gas-fired furnace. Strong's new system sidesteps the wasteful elec-

tricity stage entirely. Gas is burned to make vapor, which is used to spin a turbine, which in turn runs the heat pump.

"Fuel with external combustion is thus converted directly into mechanical power," Strong explained. "So the effective use of primary energy for heating can be doubled. Our overall thermal efficiency is therefore twice that of a conventional boiler."

Strong is the senior projects engineer at the Glynwed Central Resources Unit in Solihull, England. In his labs, I saw an impressive test rig in operation, and a prototype that could be on the market by 1985.

His prototype domestic boiler puts out 11 kW of heat energy, derived equally from the pump and Rankine vapor cycles. Output could be easily increased by increasing the turbine's and the heat exchanger's size.

Aside from its efficiency and energy conservation, the system is silent, virtually maintenance-free, and could use any fuel: gas, oil, coal, biomass, or even solar radiation. No lubrication is needed because the turbine, the only critical moving part, literally floats. Angled grooves in the spinning shaft and circular thrust face pump ambient vapor from the working fluid into the minute clearance gaps;

**Breadboard rig** in Glynwed's Solihull labs allows easy access to components for testing and modification. David Strong points to turbine case housing tiny rotor that runs silently at 150,000 rpm. Data-logging console is at right.

**Three-inch turbine rotor** is only major moving part in heating system. It combines a two-stage power turbine for the Rankine cycle with a compressor for the heat-pump cycle. Frictionless gas bearings, not on this unfinished unit, are herringbone grooves on the central shaft and radial ones on the thrust face at one end. These make oil lubrication unnecessary.

the rotor is supported without any metallic contact. There's virtually no friction or wear, and no oil to contaminate the fluid.

Possible air conditioning is another plus. For cooling, the flow in the heat-pump cycle could be reversed so the evaporator and condenser swap functions. Shut off the power-cycle condenser and you've got refrigeration instead of heat.

Glynwed spent $200,000 alone on the world's smallest turbo-compressor, though it looks simple enough. On the heat-pump side there's a two-stage expansion turbine, and a radial compressor at the other end of the shaft. It's only three inches long, but packs a lot of research and development.

That know-how was needed to create a miniature turbo machine that could run

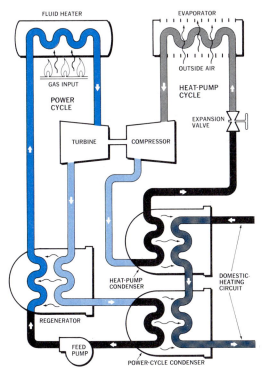

**In Rankine power cycle,** Freon-type fluid heated by gas burner expands into hot vapor through turbine driving the compressor. Most heat reaches the heat exchanger, heating the domestic-water circuit as it turns fluid. The heat pump follows a traditional vapor-compression refrigeration cycle. Chilled vapor, warmed by the outdoor evaporator, is compressed and thus heated. The hot gas condenses and is used as a second input to the water circuit. An expansion valve then lowers the fluid temperature and pressure, and discharges it to the evaporator as a vapor colder than outdoor air.
ILLUSTRATION BY GENE THOMPSON

about the rotor's gas bearings, though, since these are commonly used in dental drills running at 250,000 rpm without lubrication and are figured to have a 30-year life.

The working fluid is still being intensively studied. Thermal stability is most important, since the organic compound won't work if it's decomposed by overheating. It must also be nonflammable and nontoxic, and have antifreeze properties.

One single chemical mix must be optimized for both the Rankine and heat-pump cycles, which have slightly different

efficiently at 150,000 revs, a speed that gives the best match between blade speed and nozzle velocity. Mass producing these precision parts is one of the many problems yet to be licked. Strong has no doubts

**Domestic boiler** (shown in mock-up) will be in production by 1985. The cabinet would house all parts for the Rankine power and heat-pump cycles now at the breadboard stage. Separate evaporator at left will be fitted outside the house.

needs. Small leakages between the separate fluid circuits via the turbocompressor are then unimportant, simplifying shaft seals. Perfecting the design of the gas-fired fluid heater for this highly specialized application is another project.

## IN THE HOME

Water-outlet temperature in the prototype is a modest 60 degrees C (140 degrees F)—barely high enough for effective space heating with ordinary radiators. Strong sees fan convectors as the most suitable units for a low-temperature heating system.

These have a highly efficient surface-to-air heat transfer, and the fan motor uses no more electricity than a 30-watt light bulb. They'd work well in the cooling mode, too, as forced air reduces the condensation problem with panel radiators. Glynwed has already designed an unobtrusive convector—only four inches deep, with thermoelectronic speed control for the quiet centrifugal blower.

Heat-pump performance, using ambient air, falls off as outdoor temperature drops. To counter this, an auxiliary gas heater for the water circuit is planned as a backup in really cold weather. The future package will include microprocessor controls for economic year-round operation.

Initial cost of the new heating system would be two or three times that of a conventional boiler, Strong estimates. But you might break even in three years from fuel savings—in even less time if prices continue to rise.

# Gas heat pump

*This new pump uses a linear engine/compressor*

## By V. Elaine Smay

"This gas-fired heat pump could reduce heating bills by 60 or 70 percent compared with a conventional gas furnace," says Tom Braun, president of Tectonics Research, Inc., Minneapolis, Minnesota.

"And in summer it should provide cheaper cooling than conventional air conditioning."

The gas heat pump he refers to is not yet a reality, but Oak Ridge National Laboratory has awarded Tectonics and Honeywell a two-year, $2.1 million contract to develop it. Honeywell will manage the project and develop the controls for the system. Tectonics will adapt its combination engine/air compressor to handle a refrigerant instead of air.

In the cutaway shown, the engine is on the left and the compressor section is on the right. Between them is a counter-mov-

ing balancer to reduce vibration. Here's how it works: A spark ignites gas in the engine section and the gas expands, driving a free piston. This piston drives a second piston to compress the refrigerant and move it through the system. The pistons return to their original positions and the cycle repeats. As in an electric heat pump, in winter the refrigerant circulates through the system absorbing heat from the outdoor air via the evaporator and releasing it to the house via the condenser.

In summer, the cycle is reversed so that heat is removed from the indoor air and rejected outdoors. Waste heat from the engine/compressor also will be recovered to supplement space heating in the winter and to preheat domestic hot water in the summer.

The present goal is to develop a heat pump scaled to small commercial buildings. But the technology could be adapted for residential use with comparable efficiency, according to Braun.

# Double-duty heat pump stores chemical heat, too

*"Chemical sponge" collects solar energy to power a heat pump*

## By David Scott

If you could store summer's abundant solar energy for winter use, rooftop collectors could supply a family's year-round heating needs.

That dream of free energy was the starting point for Professor Ernst-Ake Brunberg, a physicist at Sweden's Royal Institute of Technology in Stockholm. Though practical long-term energy storage has stumped scientists for years, Brunberg thinks he has the answer: a chemical heat pump.

Like a heat pump, Brunberg's Tepidus system transfers heat (against the laws of thermodynamics) "uphill" from a cool place to a warm one. Unlike a heat pump, Tepidus doubles the energy extracted from the cool place, augmenting it with a chemical reaction that generates its own heat. And, unlike other types of chemical heat pumps, Tepidus can store energy indefinitely.

Since November 1979, the Tepidus system has been on trial in a five-room house near Stockholm with 100 square meters of floor space. Eight giant salt tanks, called accumulators, are crammed into one end of the basement. They're fed from 40 square meters of solar collector panels, which should provide for a whole year's heating needs. There is no other energy source, so no fuel bills.

A real breakthrough? Brunberg claims a good deal for the system. It has a remarkable energy-conversion efficiency of 95 percent, he says, and very high energy density compared to other storage media such as water, rocks, or phase-change salts. Brunberg also claims low losses for the system, which has few critical moving parts. And sodium sulfide ($NaS_2$), the chemical used, is cheap and needs neither replacement nor attention once installed.

This simple salt is key to the system's energy-storage cycle. Sodium sulfide is a hygroscopic salt—when it absorbs moisture, it heats up. That's what happens when you mix caustic soda with water to clean a clogged drain—the water molecules chemically combine with the salt, forming a hydride and releasing

73

**Touch-testing salt tank,** Professor Ernst-Ake Brunberg checks temperature before sealing. Insulation serves merely to retain sensible heat for immediate domestic use. When sealed, a tankful of fully dried, energy-charged salts loses only five percent of its potential energy content—even when it cools to room temperature. This means that charged salt tanks can be transported and stored indefinitely until heat is needed.

heat. Sodium sulfide has an added advantage—it won't dissolve if it's just dampened with water vapor. So a tank of damp salt can provide a bank of stored heat for warming a house and heating water.

Once the salt cools, it can be "recharged" by solar energy or waste industrial heat. Heat dries the salt, giving it the potential to reabsorb moisture and regenerate chemical heat. The vapor driven off flows to a second container, where it condenses and is held ready to wet the salt again when direct solar energy falters. This reversible cycle can be repeated indefinitely.

But Tepidus is more than an energy-stored system—thanks to the crafty method Brunberg devised to provide the water essential to the process.

The salt tank is connected to the water tank via a single pipe. The system is sealed, and the air pumped out. At this ultralow surface pressure, water vaporizes at 10 degrees C (50 degrees F)—the temperature underground. So the water tank is plugged into the ground via a heat exchanger coil.

## GROUND-WATER HEAT PUMP

"Upgrading of cold energy from the ground is the heat-pump action," says Kjell Bakken, managing director of the Tepidus company.

Because of the low pressure, there is already some water vapor present in the system. "By the nature of the salt used," adds Bakken, "if both tanks are at the same temperature, the vapor pressure on the salt is lower than the vapor pressure on the water.

"So two things happen. Because of the pressure differences, the salt wants to draw water vapor out of the water tank. That means the water needs to be evaporated. And to be evaporated it needs energy. That energy is taken from the outside—from the ground coil.

"When the vapor reaches the salt tank, it condenses. So you get back the heat taken from the ground at the same time as chemical energy is released in the hydration reaction."

The system reaches equilibrium when

**A tankful of hot salt** can store enough solar energy for year-round house heating. The hygroscopic salt stores energy when heated—releasing that energy in a chemical reaction when it absorbs water to become a hydrate. Solar-heated hot water circulates through a heat exchanger in basement salt tank (right) before traveling back up to house radiators and the hot-water tank. Basement tank at left stores the water used for the heat-releasing reaction. This tank is cooled by water pumped through a buried heat exchanger. The sealed two-tank system is kept at low pressure (0.08 psi) by an automatic vacuum pump. At this low pressure, water in the cold tank vaporizes, extracting heat energy from the ground. When the water vapor travels over to dampen the salt, it not only generates heat through chemical reaction, it also condenses, giving up the heat taken from the ground. What draws the water vapor to the salt tank? A slight drop in pressure caused by the salt cooling as its heat is tapped for house systems when the solar system isn't functioning. When the solar system *is* working, a slight rise in temperature is enough to disturb the equilibrium in the other direction. The solar heat then drives off the vapor to the cold tank where it condenses, dissipating its heat to the ground. The dehydrated salt is now rich in energy since it has the potential to once again absorb water and give off heat. With regular solar input during the summer, this hot-to-cold vapor flow continues until the salt is completely dry. A valve seals off the dry salt tank while another is charged. (ILLUSTRATION BY RAY PIOCH)

the temperature in the salt tank is 55 degrees C (131 degrees F) hotter than the water tank. Then the pressure in the two tanks is equal, and the chemical sponge no longer absorbs vapor. Disturb this

balance, and vapor is released from one tank and flows to the other. The direction of flow depends on which way the temperature swings.

Because the system is so sensitive to

minute pressure changes, the working temperature of the salt tank remains constant at about 65 degrees C. That's enough for radiators and tap water.

## ENERGY-RICH SALT

But what about long-term energy storage? Tepidus has an enormous capacity. One kilogram of sodium sulfide can store and give back one kilowatt-hour (3413 Btu) of heat. In practical terms, that means that eight tons of the dry material can store and deliver 8000 kWh (27,304,000 Btu). That's enough to meet the space- and water-heating needs of a small, well insulated house. The energy input can come from solar power alone—even assuming half the year's energy collection has to be stored for winter.

Eight one-ton salt containers, each one meter square and 1.75 meters high, make a pretty massive bulk to fit in a basement. Their volume is around eight times that of an oil tank with the equivalent thermal content. But one load of salt lasts a lifetime and soon offsets the penalty of dead space.

And Tepidus is far ahead of other forms of heat storage, according to Bakken. One main rival is eutectic salts that rely on the heat released during phase-change. "These have a capacity of about 100 watt-hours (341 Btu) per kilogram, one-tenth of our energy density," says Bakken. "And the stored energy doesn't last long because it's 'sensible' heat—perceptible to the touch—and is gradually lost to the environment.

"By contrast, ours is a low-temperature and long-term system. It can be switched off for an indefinite period and allowed to cool to room temperature. When it's started up again, only four or five percent of the total energy is used for reheating it."

And, in many installations, the storage tanks need not be so massive. The present system's capacity to store a full 50 percent of annual heating needs is calculated for Stockholm's latitude. At lower latitudes, solar input is more evenly spread around the seasons, so the solar collectors could take more of the direct heating load.

## PORTABLE HEAT ENERGY

The system opens up exciting possibilities. Energy input could come from waste heat produced in factories, such as paper mills and steel plants, as well as from sunshine.

Portable heat is a further possibility. Since the energy potential is trapped in the dry salt when it is fully charged, the sealed tanks can be disconnected and moved for hookup to a distant system as an instant heat source. Brunberg's company is designing salt-tank cargo containers for standard flatbed trucks. Surplus heat from nuclear generating stations and other plants could be delivered to remote areas in this way. Travel would be completely safe, since the tanks contain nothing explosive, and the dry salt has neither high temperature nor high pressure.

And since salt-tank modules can be any size, they could be scaled down for campers or pleasure boats. The salt could be charged by waste heat from the engine or at special heat-input stations at camp sites or marinas.

Finally, the system could also be used for air conditioning. Since the 55-degrees-C differential can be locked almost anywhere on the temperature scale, the heat exchanger for the water tank could be placed in the room to be cooled or in a cold box. Air in the cold box would hover near the freezing point as heat was extracted by evaporation inside the tank.

What are the snags? "We've fully

proved the principle in our pilot plants,'' Bakken says. "Due to the late installation, long after the sun had left Scandinavia, we had to charge the system with resistance heating,'' he adds. "But it was fully loaded in February, and we heated the house from then on. Now we're charging the tanks again for winter.''

## SECOND TESTING SYSTEM

Bakken also reports that another pilot system, with a storage capacity of about 30,000 kWh (102,390,000 Btu), has recently begun operating.

"The critical salt material has given no trouble,'' he says. "We've run 150 charge-discharge cycles already—which could equal 150 years' normal operation—and the salt has not deteriorated.

But accelerated cycling isn't good enough, as it may give spurious results. Only when we are able to test complete systems for 30 years can we say there are no problems.

"Our goal now is to improve component design to cut production costs. There are no dark areas, only problems of technical detail. Heat exchangers, for example, become very expensive when you're handling lots of power.''

Initial cost could be a problem. Bakken figures a complete Tepidus installation for an average family house would cost four or five times as much as an oil burner. But he notes that the solar collectors would be a major item. "And remember we're talking about a one-time investment that you have to balance against future bills.'' That balance could favor Tepidus as fossil fuels become costlier.

# 3 WHAT ABOUT YOUR THERMOSTAT?

**W**hat about your thermostat? That's a good question, for it plays a very important part in the operation and effectiveness of your home's heating system—in fact, you can think of it as the "brains" of the heating system.

Unfortunately, the thermostat is too often taken for granted. Don't make that mistake! In a few minutes time you can give it the care it needs to operate your heating plant economically, and often improve comfort as well. And with one of the new-generation electronic thermostats, you can even raise your system's "IQ" to a much higher level.

# Computerized thermostats program your comfort *and* save on energy costs

*These sophisticated electronic wizards can put your home on a customized energy diet*

## By Richard Stepler

As energy costs increase, it's vital to have more precise control over home heating and cooling systems. Now the revolution in microelectronics has produced the computerized thermostat, an instrument that promises to keep your home comfortable and cut heating and cooling costs by up to 30 percent.

What can these new microprocessor-based gadgets do that your present thermostat can't?

- Automatically provide two, three, or more setback and setup temperature periods a day.
- Provide a different schedule of times and temperatures for weekdays and weekends.
- Digitally show time, day, and actual and desired temperatures.
- Let you override the setback/setup program by flipping a switch.
- Plus, one of them gives temperatures in Fahrenheit or Celsius, and another even displays outdoor temperature.

A number of companies have introduced computerized thermostats, ranging from giant Honeywell, the thermostat pioneer, to newcomers in the field, including Texas Instruments. The survey also revealed controversies: Are the thermostats too complex? How should they be powered? How accurate are they?

"While engineers may drool over the potential, consumers will make the purchasing decisions," Art Rotman, manager of Honeywell's residential heating and air-conditioning markets, told me recently. He was explaining why Honeywell decided to introduce a hybrid: a thermostat with conventional bimetal temperature-sensing hardware, coupled to a digital electronic clock. With a little prompting from Rotman, I found I could easily program two setback periods into the thermostat's memory, and choose a setback temperature. A switch selects either a seven-day or five-day program; choosing the latter eliminates the day setback on weekends.

Micro Display Systems and Johnson Controls avoid keyboard controls altogether, instead using "more intuitive" slider controls to set times and temperatures (see photos). "It's even easier to read than conventional thermostats," boasts MDS president Tom Hyltin. "You can walk up to it and tell it what to do, never having seen it before."

RapidCircuit's electronic thermostat also uses levers to set desired temperatures, which are shown digitally on the unit's liquid-crystal display (LCD).

Other thermostats, such as Autotronics' and Control Pak's, offer more complex programming capabilities, but require more knowledge to adjust. They should discourage thermostat tampering, however. (The Autotronics thermostat even has a lock.)

The Control Pak and Autotronics thermostats use a separate low-voltage transformer, usually installed in the furnace room, as a power supply. This, they say, is safer than the thermostats that draw their power through the furnace's low-voltage wiring. On gas furnaces, a valve or flue damper might be inadvertently actuated, they claim, and asphyxiation or an explosion could result.

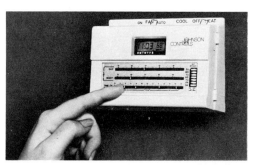

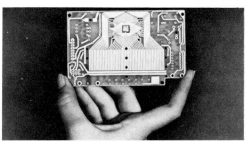

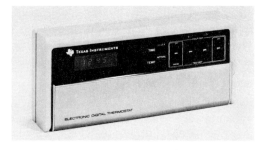

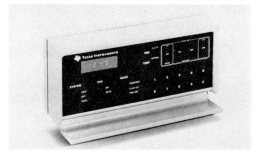

**Johnson Controls microelectronic** thermostat sets back during the day and night. Detented sliders control times and temperatures. Day-of-week switches set for days when house is normally occupied override day setbacks. LCD shows time, temperature, day, setback/setup mode. Microprocessor (dark square in center) is shown in view of the unit with its cover removed: Basic unit sells for around $120.

**You program Texas Instruments'** thermostat on a keyboard for up to four different time/temperature settings per day. Flipping a switch changes unit to a weekend program, or to a preset constant temperature. LED display shows time and temperature. Unit operates from 24-volt thermostat wiring or nine-volt battery that acts as backup in case of power failure. Price about $125.

The MDS thermostat operates on very little current—2½ milliwatts—and literally steals it from existing low-voltage wiring, even when the furnace is actuated. It's too little, claims Hyltin, to cause the potential problems cited by Autotronics and Control Pak. The Honeywell, Texas Instruments, and PSG Industries thermostats also are powered by existing low-voltage wiring, but draw more power—50 milliwatts to one watt. These incorporate a battery to act as a backup power source in case of power failure.

How a thermostat senses temperature and controls the operation of a furnace is not quite as simple as you might expect. Different heating systems have different response times (fastest: electric forced air; slowest: hydronic). In the case of slow-response-time systems, the thermostat must be able to turn the furnace off while room temperature is still rising to prevent temperature overrun.

Many conventional thermostats, including Honeywell's computerized T800, employ a mercury capsule switch, bimetal coil, and an anticipator—a tiny heat source—to maintain temperature close to the setpoint.

PSG's Compustat uses preset mercury sensors to maintain temperature within 2° of the thermostat's setting, the company claims.

MDS's computerized thermostat senses temperature electronically with a diode. "We're doing things that a mechanical thermostat can't do," claims Hyltin. The MDS unit, he says, is a second-generation thermostat with an electronic anticipator that modifies the setpoint as a function of time. "It works with any heating system—from electric to hydronic—with no adjustment," he says.

RapidCircuit's model operates on a temperature differential. "Set at 68° F," explains engineer David Sandelman, "the thermostat will allow room temperature to rise to 70° before shutting the furnace off, and down to 66° before turning it

**Honeywell's T800 also uses keyboard** for programming; instructions are on back of flip-down cover plate. It's capable of two setback/setup periods (up to 15°F) a day for either five or seven days. LCD gives time of day or setback times only. Price, says Honeywell, is around $200. This thermostat uses bimetal hardware to sense temperature; most others sense it electronically.

back on." You can adjust the differential from ½° to 15° to allow the house to heat evenly. "How you adjust it," explains Sandelman, "depends on the type of heating system, size of your house, and how well insulated it is."

Control Pak's Temp-Tron uses forced cycling to minimize droop (average temperature below setpoint over a given time). The installer selects either three cycles per hour (slow-reacting hydronic) or six cycles per hour (fast-reacting forced air) of furnace on-time. The length of the cycles is varied; the greater the load, the longer the furnace is on.

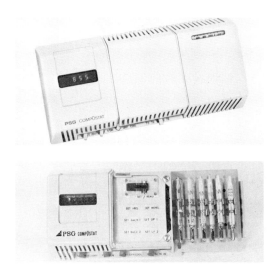

**Preset mercury sensors** (on right in lower photo) in PSG's Compustat provide temperatures from 68 to 78° F in 2° increments, plus 60° setback for heating and 80° setup for cooling. Two setback/setup periods a day can be programmed on keyboard. LED display gives time of day and setback/setup times. Heating only model: about $80; heating/cooling: about $100.

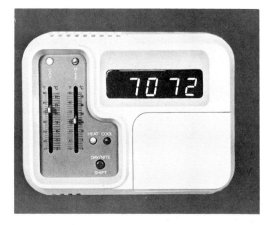

**RapidCircuit's electronic thermostat** shows desired and actual temperatures on half-inch-high LCD, which also gives time of day and setback/setup times. Two low-voltage wires power it; there's no battery. Integrated circuit senses temperature. Heat-only single setback model costs approximately $99; dual setback unit about $109. Heating/air conditioning models are slightly higher.

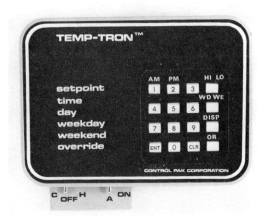

**Temp-Tron from Control Pak** can make up to eight changes per 24-hour period in two different schedules. LED display shows desired and actual temperatures, or time, day, and outdoor temperature via remote sensor.

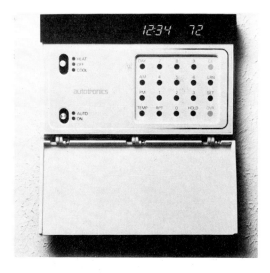

**Program up to 28** different time and temperature settings a week (four per day) on Autospace 7's keyboard. Autotronics unit has continuous readout of time and actual temperature on LED display. Override key lets you interrupt preset program without erasing it. It replaces four-wire thermostats. Separate low-voltage transformer powers this thermostat and one shown at left.

## How much can a setback/setup program save you?

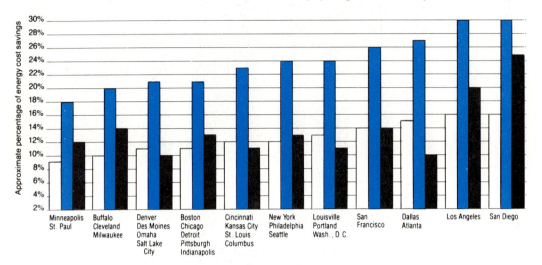

**Honeywell charted** these energy savings for 29 cities. Heating: 10° F eight-hour night setback from 70° (left bar); 10° F eight-hour night and day setbacks from 70° (middle bar). Cooling: 5° F 11-hour day setup from 75° (right bar).

## PROGRAMMING THE THERMOSTAT

Here's an example of how one of the computerized thermostats might be programmed for a typical weekday (for weekends, of course, you could program a different schedule). At 6:30 AM, the thermostat actuates the furnace to raise house temperature from 58 to 68° F so you'll be comfortable when you awaken at 7. At 8:30, just before the last member of the family leaves for the day, the thermostat sets back to 58°. At 4:30 PM, it goes up again to 68° so the house is comfortable by 5. At 10:30 PM, when you retire, it sets back again to 58°. Downtime: 16 hours.

It should be news to no one that setting your thermostat down in the heating season and up in the cooling season will reduce your energy costs. In fact, depending on where you live, you could save up to 30 percent in winter and 25 percent in summer, according to a Honeywell study (see chart). Honeywell based these figures on 10 years of weather tapes from 29 cities and test results from a house occupied by a family of four. You can see that the really big savings for heating come when you have two setback periods—a total of 16 hours, as in the example above.

Like first-generation pocket calculators, the electronic thermostats are costly, with prices ranging from just under $100 to more than $200. Conventional clock thermostats, which are capable of up to two setbacks, are typically priced from $40 to $60; you'll have to decide if the digitals' added features are worth the premium price. Note, though, that all setback thermostats presently qualify for a 15-percent federal income-tax credit.

Prices of the electronic models will undoubtedly come down, however. "Obviously, electronic thermostats will go through a price curve similar to other electronic components," predicts MDS's Hyltin. RapidCircuit, which claims to have introduced the world's first such thermostat now offers a $99 model. That's half the price of the company's original.

# Installing a super-smart setback thermostat— even on a heat pump

*It's programmable for specific days of the week*

## By Evan Powell

No question about it: An automatic setback thermostat can save you energy. Newer models, with up to four setback times, are more flexible and convenient. And now there's SmartStat 1000, which even works with a heat pump.

It's programmable. A keyboard lets you enter all your heating and cooling requirements at one time. It accepts day and night temperatures for specific days of the week. Take a trip, for example, and it goes into its setback mode as you walk out the door. On the day and hour you return, SmartStat has readjusted the temperature to normal.

During a setback, SmartStat monitors the frequency and length of running time required to maintain the low temperature. This calibrates the unit so it knows precisely when to turn on the system to reach the temperature you've set at the required time. The feature makes it perfect for use with a heat pump since no resistance heating is needed. SmartStat will simply start the compressor earlier after a very cold night than it would after a mild one.

The SmartStat replaces your present thermostat, but an additional wire must be added down to the furnace. Also, a supplied relay module must be added to the existing wiring. This connects the SmartStat control circuits to your furnace.

The unit has a built-in temperature sensor, but an optional remote sensor may be added for sampling air from other areas. This allows the unit, with its alternating time/temperature display, to be conveniently located, while its actual temperature sensor is near an intake duct for a more accurate air sample.

Made by NSI, Sugar Hollow Rd., Morristown, Tennessee 37814, SmartStat is also marketed nationally by Johnson Controls.

**Relay panel** above mounts on furnace, connects in line to the existing system. Optional temperature sensor (above right) measures air samples at remote locations.

# Tune your thermostat to save heating dollars

*Improve an existing unit or use an auto-setback stat—either way, you're ahead*

## By Evan Powell

That little instrument on your wall called the thermostat has just one job—it must turn the furnace burner on and off to maintain a preset temperature. That's a simple task, but many thermostats don't do it well. And when they don't, you pay—operating costs are high and the comfort level is low.

Thermostat malfunction is usually due to one of three causes: The device isn't located, maintained, or used correctly. But chances are good that you can easily solve these problems yourself. And if you'd prefer to replace the unit, there are a variety of new controls that offer greater fuel savings and don't require extensive modification or rewiring.

Before you tackle a maintenance or a complete replacement job, however, start from the beginning:

## STAT LOCATION

Too often, little thought has gone into the original placement of your thermostat. Or remodeling or furniture rearrangement may have made the present location unsuitable. A thermostat should detect the *average* temperature in your living area. It shouldn't be influenced by sudden blasts of cold because it's located near an outside door or mounted on an exterior wall. Heat—from lamps, supply registers, and TV sets—will also trick the thermostat into wasting fuel or affecting your comfort level.

Often you can remedy such a problem by relocating a room's furnishings to remove a heat source or shield the stat from cold drafts. But if you move the location, consider buying a new thermostat. Disconnect the old one, but leave it in place until next time you paint or paper the room. Naturally, you'll have to pull low-voltage wiring to the new location.

## TUNING YOUR THERMOSTAT

Although makers recommend only an occasional light dusting of the thermostat components, the additional adjustments I often find necessary don't violate the manufacturer's chief concern: tampering with or recalibrating the unit. They've had bad experiences with bent, deformed bimetal springs and serious miscalibration resulting from improper repair techniques.

Fact is, the best maintenance is simply a check for proper installation and performance—no less than you would do if installing a new unit from scratch.

To operate properly, a thermostat—particularly one with a mercury-bulb switch—must be absolutely level. Often, a thermostat has not been leveled properly when first installed. Or the mounting screws may have loosened, particularly in plaster walls. Or the thermostat may have received a bump hard enough to knock it off level.

To check level, turn off all power to the furnace and remove the front cover or trim ring of the stat. This usually snaps in place, and an easy pull should loosen it. Look near the top edges of the unit for a line marked "level" or for two plastic pins with flat areas on their top side. On round thermostats you may have to remove the front assembly, which is held to a baseplate with three recessed screws (see photo).

Place a small spirit level across the leveling "ears" and adjust the thermostat body or baseplate by loosening the mounting screws and turning the thermostat base within the elongated mounting slots until the control is level. Once it's level, reinstall all the components except the cover.

## SET THE ANTICIPATOR

You will find the anticipator adjustment on the thermostat mechanism and marked with a series of numbers. At opposite ends of the scale you'll usually find the words

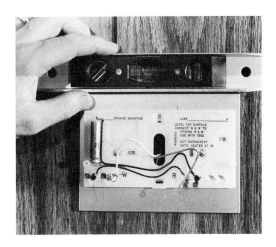

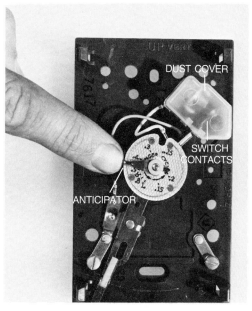

**Mercury-switch thermostats** must be level (upper photo). If mechanism must be removed, be sure hardware is tight when reinstalling—it provides electrical connection to stat. Set anticipator, above, to match furnace rating.

"shorter" and "longer." These apply to the length of the burner running time within a heating cycle, but the initial setting should correspond with the rating of the thermostat circuit for your furnace.

If you have the owner's manual for your heating plant, this figure should be listed as "low-voltage secondary circuit current" or "thermostat circuit current." It will be expressed in amperes, such as "4.5 amps." Then you simply set the pointer of the anticipator to 4.5. If you don't have this information, look at the primary control or gas valve on your furnace; it should be printed on one or the other. You can recognize these components by tracing the small low-voltage thermostat wiring to the point where it is attached to the furnace. It may be necessary to turn off the power and remove the outer cover of the control box to read the information.

When you have made this initial setting, you may make finer adjustments if you notice a great variation in room temperature between heating cycles. If the room is warmer than you like while the furnace runs, reduce the anticipator setting slightly toward the shorter end of the scale, and do the opposite to increase temperature. Wait at least 48 hours after making any adjustment before passing judgment, since it takes time to level off. Ideally, you will now notice very little variation in temperature within the room, whether or not the furnace is running.

## How a thermostat works

The thermostat is the brain of your heating system. It must swiftly detect minute changes in room temperature and turn the heat source on and off accordingly. Moreover, it has to anticipate "overrun" of the heating system—the period when the blower of a warm-air system or the pump of a hot water-system continues to operate after the burner has shut down.

The thermostat is basically a low-voltage, temperature-sensitive switch that opens and closes its contacts to control a relay or solenoid at the heating plant. The switch in many thermostats is a glass capsule filled with mercury. Wire fingers, protruding into one end, serve as the contacts, and the capsule is attached to a bimetal coil. The two metals have different coefficients of expansion, causing the coil to bend in response to temperature change. As temperature in the room drops, the coil movement causes the capsule to tip, and the mercury flows around the "fingers," closing the circuit.

In other thermostats this job is done by a bimetal strip with a contact at the end. (Often a small magnet is used to provide "snap action" of the contact.) As room temperature warps the strip, it opens and closes the contact. The starting position of the bimetal spring or arm (and therefore its actuating temperature) is physically altered when you set the dial that selects temperature.

Another part of the thermostat is the anticipator circuit, which provides for blower overrun. The blower in most warm-air systems is not controlled directly by the thermostat, but by a temperature-sensing switch within the furnace. This prevents the blower from coming on while the furnace is still cold, or shutting off immediately while the heat exchanger is still hot and capable of supplying heat.

To make allowance for this delay, the thermostat contains a tiny electric resistance heater that causes the thermostat to open the burner circuit a bit early, maintaining room temperature close to thermostat setting during overrun.

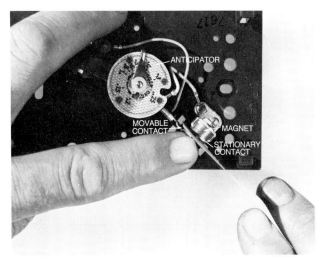
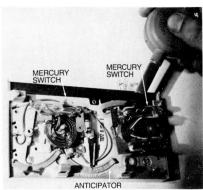

**Clean contacts** (as shown, left) in the closed position with a piece of white bond paper. Photo bulb-brush removes dirt and dust elsewhere (right). Use care, however, when cleaning around any delicate bimetal coil.

## THERMOMETER CALIBRATION

Very often the thermometer (usually part of the outer cover) and the thermostat settings don't jibe. Once the thermostat is perfectly level, the disagreement usually is the fault of the thermometer. (Don't attempt to recalibrate the thermostat. This requires special tools and is almost never necessary.)

To make settings coincide, with power still off to the thermostat, carefully turn the adjusting knob from a low room temperature to a higher one while watching the mercury bulb or the switch contacts. As soon as those contacts close, stop your motion and note the temperature indication on the adjusting knob. Now set the thermometer to coincide with this setting. This can usually be done from inside the thermostat cover.

Naturally, before you make this calibration, avoid touching the cover in the vicinity of the thermometer or the body of the thermostat for at least five minutes. Your body heat can influence the setting by several degrees.

Many thermostats use open contacts for control. These contact surfaces should be cleaned by pulling a strip of white bond paper between them while they are closed. A tiny particle of dust can keep them from operating.

Clean the on/off switch and the other contacts in the same way. Finally, using a soft vacuum brush or small photo brush with a blower bulb, carefully clean away any dust from within the thermostat mechanism. Check wiring connections to be sure that they are secure, and replace the cover.

## PROPER USE—AND NEW STATS

Find the lowest temperature level at which you can set your thermostat and still keep comfortable in all areas of your home. Then leave it there. Don't be a "thermostat twiddler." The stat can't make the house heat faster, so never set it above the level that you want to maintain.

It has been positively established that nighttime setback is one of the best methods of reducing fuel consumption. There are various new products and do-it-yourself projects to accomplish this, and commercially available clock thermostats have been on the market for many years.

Several new products will give you automatic control without wiring modifications, since they "borrow" from the low-voltage circuit of the furnace or use battery power. I have installed and used several types and can vouch for the fact that some will pay for themselves within two years. Many are marketed for do-it-yourself installation.

**Savings are automatic** with setback thermostat like this Honeywell model that can be set to control temperature setback at night.

# Thermostat setback

*Here's proof it saves*

## By Ed Moslander

Depending on where you live, regional heating-fuel prices, your house size, and the quality of its insulation, you can save up to $200 annually on fuel by setting back your thermostat at night and when no one is home during the day.

Fuel savings from thermostat setbacks have long been accepted as fact, but little documentation existed to support it. Now a study, conducted in 38 cities in the nine most populous cold-weather states (all in the Northeast and Midwest), appears to supply that documentation. It used winter-1980 heating-fuel prices.

The study was conducted by Honeywell, a leading manufacturer of thermostat setback devices. It says turning down your thermostat from 70 to 60 degrees, for two eight-hour periods each day, saves 14 to 21 percent per year in heating costs. If someone is home all day, a single setback of 10 degrees for eight hours at night saves between eight and 13 percent. If a 10-degree setback is too severe, turning down your thermostat five degrees for two eight-hour stints produces savings of between eight and 15 percent. And even if you reduce your setting only five degrees for one eight-hour span, you still save between four and nine percent annually.

The most reliable and convenient way to achieve these energy and money savings is to install an automatic-setback or clock thermostat, Honeywell says.

A clock thermostat can be programmed to make four adjustments daily, turning the heat off when you're sleeping and when no one is home, and turning it on again before you get up in the morning and arrive home in the evening. You don't need to reset the heat, nor do you wake up to a cold house in the morning.

Installing a clock thermostat costs between $60 and $150. But the initial investment can pay for itself within one to three years, Honeywell says. For example, if you live in New York City and heat your home with oil, setting back your thermostat 10 degrees for two eight-hour periods saves $167 a year (see table). If you live in Chicago you save $180, and in Albany, N.Y., $199, the largest saving of any city studied.

If you heat with natural gas, the repayment period is longer. But with the price of all heating fuels expected to rise, the savings should get larger and the payback period shorter.

In addition, the federal government allows a tax credit of 15 percent of the purchase and installation costs, further reducing payback time.

## Average annual saving: GAS and OIL

| City | Saving with single setback (%) 5°/10° | Saving with dual setback (%) 5°/10° |
|---|---|---|
| Albany, N.Y. | 7/11 | 13/19 |
| Boston, Mass. | 7/11 | 13/19 |
| Buffalo, N.Y. | 6/10 | 11/17 |
| Chicago, Ill. | 7/11 | 12/18 |
| Cincinnati, Ohio | 8/12 | 14/21 |
| Cleveland, Ohio | 7/10 | 12/18 |
| Detroit, Mich. | 7/11 | 12/18 |
| East St. Louis, Ill. | 9/13 | 15/21 |
| Grand Rapids, Mich. | 6/10 | 11/17 |
| Green Bay, Wis. | 5/9 | 9/15 |
| Harrisburg, Pa. | 8/12 | 14/21 |
| Indianapolis, Ind. | 7/11 | 13/19 |
| Lansing, Mich. | 6/10 | 11/17 |
| Madison, Wis. | 6/10 | 10/16 |
| Milwaukee, Wis. | 6/10 | 10/16 |
| Newark, N.J. | 8/12 | 14/21 |
| New York, N.Y. | 8/12 | 14/20 |
| Philadelphia, Pa. | 8/12 | 14/21 |
| Pittsburgh, Pa. | 7/11 | 13/19 |
| Rochester, N.Y. | 7/11 | 13/19 |
| South Bend, Ind. | 7/11 | 12/18 |
| Springfield, Ill. | 7/11 | 13/19 |
| Syracuse, N.Y. | 7/11 | 13/19 |
| Trenton, N.J. | 8/12 | 14/21 |

# Light-sensing thermostat for electric baseboard heat

## By Evan Powell

The Li/Tronic thermostat takes the place of conventional room thermostats used in zonal electric heat systems. But it doesn't work conventionally.

In daytime, or when the lights are on in the room, the stat operates at a preset high level. But when the room gets dark—when you go to bed—it automatically drops eight to ten degrees, saving energy and money.

The unit differs from other light-sensing thermostats in that it is designed to handle the higher power levels used in zonal electric systems. It's the first practical setback thermostat for electric baseboard heat and radiant heating systems with a stat located in each room. No additional wiring is necessary—the Li/Tronic simply replaces your present stat.

Breakaway ears on the control knob determine the high and low limits of the range selection; a "normal" switch overrides the setback feature. The unit uses a bimetal temperature sensor with built-in anticipator. A thermometer and upper-

temperature-limit kit are also available. Made by Federal Pacific Electric (150 Ave. L, Newark, New Jersey 07101), Li/Tronic costs around $35 at retail outlets.

 # FIREPLACES

Whenever we think of the comfort and coziness our home offers in winter, somewhere in the corner of that picture in our imagination is the warm glow of a fireplace. But the typical masonry fireplace only masquerades as a heating source, because it's often a negative factor—actually taking more heated air from the home than it puts back into it. At best, it provides only a small amount of heat, not nearly enough to justify anything but the atmosphere it creates.

But you can preserve that special atmosphere of the cheery open fire and have your heat, too. In the pages that follow, you'll see how.

**Bi-fold glass doors** minimize heat loss in Heatilator fireplace. Doors are also available for retrofitting.

**Room air circulates** through 14 feet of convoluted steel heat exchanger in Majestic's Energy Saving Fireplace, channeling nearly 40 percent of the heat into the house through side grilles, claims the maker. With optional duct kits and fans, heated air can be diverted to adjacent rooms, or even upstairs.

98

# Heat-saving fireplaces

*Enjoy the flames without the waste*

## By Darrell Huff

Beset by charges that wood-burning fire-places are hopelessly inefficient, waste-ful, and even counterproductive, the people that make 'em are fighting back. As well they might, with their existence at stake. In some parts of the country, fire-places have actually been outlawed in new construction. The reason is that a conventional fireplace ineptly used may not only produce very little heat; it may even chill your house. The threat to fireplace builders is a threat to all us open-fire buffs as well.

To keep their product viable in an energy-short world, innovative fireplace makers have attacked the problem on four major fronts. Generally, these new designs are available in complete, new fireplaces. Some of the methods used, however, can be adapted to many older fireplaces.

## DAMPERS AND DOORS

The most serious charge against the ordinary fireplace is that it steals room air—air that you've spent energy dollars to heat—and chutes it wastefully up the chimney. And it may keep right on doing so even when it is cold and dead. This between-fires heat loss can total more than the heat gained during the hours a fire is burning.

All you have to do to reduce this loss markedly is close the damper as soon as the fire is fully out. That's if your fire-place has a damper—one that's not too difficult to use—and it seals effectively. And if you remember every time.

To answer all these problems, fireplace manufacturers are now supply-ing more effective and convenient dampers. Such dampers also save energy by providing fire control. A damper that may need to be wide open while a fire is being kindled can soon be partly closed to control burning speed, thus saving wood.

But there is a far more effective control available for use either alone or in con-junction with a damper: metal or glass doors to close off the fireplace opening. The prime time for using doors is when you leave the house or go to bed before the fire dies out. Obviously, you can't close the damper under such circum-stances. But you can seal off the fireplace opening by closing the doors; thus you keep the warm room air from rushing up the chimney as the fire smoulders and dies.

Some doors may be closed even when there's a fire burning. Your fireplace then converts to stove efficiency. If the doors are glass, you still have the pleasure of

99

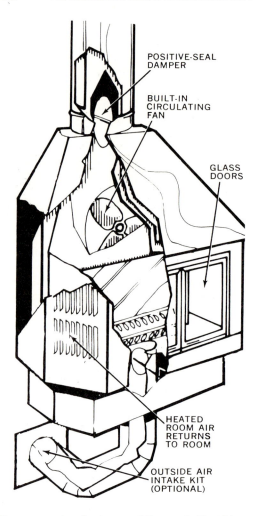

POSITIVE-SEAL
DAMPER

BUILT-IN
CIRCULATING
FAN

GLASS
DOORS

HEATED
ROOM AIR
RETURNS
TO ROOM

OUTSIDE AIR
INTAKE KIT
(OPTIONAL)

**Energy-saving features** of Preway's Provider include damper, circulating fan and grilles, glass doors. Kit to draw outside combustion air is optional.

**Malm Fireplaces' "Carousel"** produces swirling fire and radiates heat through glass walls and metal top. Closing glass door controls draft and also prevents loss of room air when fireplace is not in use.

## CIRCULATING FIREPLACES

One version of the fireplace that heats and circulates air has been around for a long time—the double-walled steel firebox around which a masonry fireplace is built. It is best known under the trade name Heatilator, used by a principal maker.

Good though it is, the circulator core had suffered a decline because of competition from the far cheaper, more adaptable, and faster-to-install zero-clearance units. But the early zero-clearance fireplaces fell short in efficiency. Now manufacturers have introduced new air-circulating versions. Air is drawn in through a slit below the firebox or perhaps at the sides, and is forced out at the top of the opening, through wall grilles above the fireplace, or to the sides.

watching the fire. A fireplace door can be an important safety device as well, protecting against sparks and rolling logs.

Among freestanding units, the equivalent of the fireplace with a glass door is a cone-shaped model called Carousel. Made by Malm Fireplaces, it is glass all the way around under a metal hood. One glass section opens for log placement and draft control.

The circulator zero-clearance units can be installed in a matter of hours directly on wood floors and against framing studs. Fire-code requirements are met by providing a hearth and surrounding the opening with a fireproof material.

While a zero-clearance circulator can go almost anywhere, the masonry-surround circulators require solid foundations to hold masonry walls. Chimneys are commonly masonry, too.

Water pipes passing through a fireplace have long been used by ingenious builders to heat domestic water supplies. In an extension of this idea, one fireplace (called Hydroplace) came with coils through which water (or other liquid) passes and conveys heat to a storage tank through a closed-loop heat exchanger. This heated water then warms the house, and, if desired, the domestic hot-water supply. The natural complement of this system is a solar heating system, for which the Hydroplace provides backup heat on dulls days. Properly sized and located pressure-temperature relief valves must always be used in such systems.

Air circulators are also available for retrofitting. Some types are simply hollow grates that sit in the fireplace. Others use fans. For freestanding fireplaces there are heat extractors that install around the lower part of the smokestack.

## DUCTING TO OTHER ROOMS

One drawback to a fireplace is that most of them heat one room but leave the rest of the house chilly. If this is your problem, you'll want a circulator designed to duct heat to other rooms. Both the zero-clearance types and the ones requiring masonry surrounds can do this. Ducting is simplest, naturally, when the second room to be heated is directly behind the fireplace. A fireplace in such an interior location will also return to the house a greater portion of the heat produced in it.

In new construction or extensive remodeling, it's usually feasible to carry ducts to more distant rooms. Some models can be ducted to rooms on other floors. The most efficient arrangement is one in which the furnace ducts are connected to the circulator fireplace as well. Any fireplace that can be connected to ducts can be used in this way. Some, such as the Pyroplace, are designed for this use.

## OUTSIDE AIR FOR COMBUSTION

Most fireplaces take their combustion air from the room they're in. Such a large quantity of air is involved that in many modern, tightly constructed houses a window must be opened for the fire to burn without smoking.

When this great volume of room air has already been heated by a furnace, it is wasteful to burn it up. Some new fireplaces include a connection to outside air. Other models have optional kits for this.

**Designed to duct heat** to other rooms, the Hot One warms room air from the base of the firebox to well up along the flue. Fan, glass doors, and external combustion air increase efficiency.

# Energy-efficient fireplaces you can buy

| Source | Model | Type | Heat circulator | Blower fan | Ducts to other rooms | Firebox has doors | Outside air for combustion | Remarks |
|---|---|---|---|---|---|---|---|---|
| **Energy Savers Co.,** Box 111, Salem IL 62881 | The Hot One | Masonry shell | Yes | Yes | Yes | Glass | Yes | Finned heat exchanger in flue area as well as around firebox |
| **Heatilator Fireplace, Div. of Vega Ind., Inc.,** Mt. Pleasant IA 52641 | Model 3138 (123C) | Zero-clearance built-in | Yes | Optional | Yes | Glass bi-fold optional | Optional | Glass doors available for retrofit |
|  | Mark C | Masonry shell | Yes | Optional | Yes | No | No |  |
| **The Majestic Co.,** 245 Erie St., Huntington IN 46750 | Energy Saving Fireplace ESF-11 Peabody | Zero-clearance built-in | Yes | Optional | Yes | Glass bi-fold | Kit with damper available |  |
|  |  | Zero-clearance fireplace-stove | Yes | Yes | No | Tempered glass on 3 sides | Preheated secondary air source |  |
| **Malm Fireplaces, Inc.,** 368 Yolanda Ave., Santa Rosa CA 95404 | Royal Franklin | Free-standing | Yes | Yes | Yes | Glass | No |  |
|  | Carousel | Free-standing | No | No | No | Glass, adjustable for draft | No |  |
| **Marco Mfg., Inc.,** 2520 Industry Way, Lynwood CA 90262 | Futura DF-36B and Monarch DF-46 | Zero-clearance built-in | Yes | No | No | No | No | 5-position positive-seal damper |
| **Marketing Services Dept.,** Box 1280, Redwood City CA 94064 | Woodside | Zero-clearance built-in | Yes | No | No | Glass optional | No | 3 sizes plus right- and left-hand corner models |
| **Martin Ind.,** Box 128, Florence AL 35630 | Octa-Therm, Quadra-Therm | Zero-clearance built-in | Yes | Optional | Including upstairs | No | Yes | Combustion air through vertical slots on sides |
| **Montgomery Ward** catalog or order office | Ward's built-in | Zero-clearance | Yes | No | Optional | No | Yes | Duct to outside air can be shut |
|  | Ward's circulator | Free-standing | Yes | Optional | No | Glass bi-fold | No |  |
| **J.C. Penney Co., Inc.,** 11800 W. Burleigh St., Milwaukee WI 53263 | Penney's Franklin-style heat circulator | Free-standing | Yes | Optional | No | Cast iron; open to view fire | No |  |
|  | Penney's Heater/ Fireplace | Free-standing | Yes | No | No | Steel; open to view fire | No | Colors: black, white, burnt orange |

| Source | Model | Type | Heat circulator | Blower fan | Ducts to other rooms | Firebox has doors | Outside air for combustion | Remarks |
|---|---|---|---|---|---|---|---|---|
| **Pre-way, Inc.**, Wisconsin Rapids WI 54494 | Custom built-in | Zero-clear-ance | Yes | Optional | No | Glass | Yes | Glass doors and outside air ducts included on Energy Mizer model only Colors: black, red |
| | Provider FB-24F | Free-standing | Yes | Yes | No | Glass | Included on Energy Mizer model only | |
| | Continental FBG24 | Free-standing | No | No | No | Optional | Optional | Available in black only |
| **Pyrosolar Ind., Inc.**, 65 Cedar Grove Rd., Rolla MO 65401 | Pyroplace | Built-in | Yes | Yes | Uses furnace ducts | Glass | Yes | Requires concrete floor or base, hearth, 4" wall clearance |
| **Sears, Roebuck**, catalog or order office | Sears built-in | Zero-clear-ance | Yes | Optional | No | Glass (optional) on some models | No | |
| | Sears free-standing | Free-standing | Yes | 3-speed | No | No | Optional kit | |
| **Superior Fireplace Co.**, 4325 Artesia Ave., Fullerton CA 92633 | E-Z Heat | Zero-clear-ance built-in | Yes | Optional | Yes | No | No | |
| | Heatform | Masonry shell | Yes | Optional | Yes | No | No | Fire-A-Lator is low-er-cost circulator |
| **Temtex Prod., Inc.**, Box 1184, Nashville TN 37202 | Temco | Zero-clear-ance built-in | No | No | No | No | No | Energy-conserving damper; special air-flow chimney |
| **Wells Fireplace Furnaces**, Box 7079, Tucson AZ 85725 | Wells Fireplace Furnace | Built-in | Yes | No | No | Glass | Yes | Intended to heat the entire house |

With a masonry fireplace it may be possible to supply outdoor air without special ducting. The ashpit can be cleaned and kept open, allowing air to be drawn directly into the firebox, or an outside air intake can be located to the side or front of the fireplace opening.

The necessity of supplying ample air to a fireplace was demonstrated to the family of Richard Sanford, a carpenter in Monticello, Iowa. Sanford had insulated to the hilt. "My house was just about airtight," he says. The family retired one chilly evening, leaving a wood fire to help their gas furnace heat the three-bedroom house. The fireplace fire consumed most of the oxygen in the house, causing the furnace to malfunction. Starved, it sent carbon monoxide through the house. The family barely escaped.

The incident is a useful reminder of the importance of proper operation as well as good design and installation of your fireplace. With these, it can contribute to comfort and energy-saving, without denying you the pleasure of seeing those flickering flames.

# Fireplace improvers

*These can get you more heat from
your firewood*

## By A.J. Hand

No source of home heat can match the
charm and atmosphere of a fireplace.
Unfortunately, fireplaces are notoriously
inefficient, allowing room air to escape up
the chimney and delivering little of the
heat actually produced by the fire into
your home. Fireplace efficiency can run
anywhere from about 10 percent down to
zero—or even siphon off existing heat.

A lot of effort has gone into the
development of devices to raise the efficiency level without sacrificing the
charm. Most of the earliest designs
concentrated on extracting heat from the
core of the fire. (Examples of this type
include C-shaped tubular grates, such as
the Thermograte, and radiating grates,
such as the Texas Fireframe.)

But experts like Harrison Edwards of
Norwich Labs are critical of devices that
take heat out of the fire itself, which
reduces the temperature of combustion,
making it less efficient.

"It's best to extract heat after it leaves
the core of the fire, but before it escapes
up the chimney," Edwards advises.

Here are two new fireplace improvers
that do just that:

## BOILER/RADIATOR SYSTEM

To prevent the escape of room air up the
chimney, Convecto-Pane Plus (developed
by Edwards for Bennett-Ireland, 23 State
St., Norwich, New York 13815) has a fireplace enclosure with double-glazed doors.
It seals tightly against the fireplace
opening. A built-in draft control lets you
regulate air flow into the fireplace so you
admit just enough air for proper combustion.

You can cut off the air supply completely when you retire for the night. The
fire will be safely contained as it dies
down, and no room heat can slip up the
chimney.

The enclosure primarily prevents heat
loss, but it also delivers heat. The air
space between the panes in the doors is
vented, which allows air to pass between
them, picking up heat and carrying it into
the room.

The system can deliver about 5,000 Btu
per hour, including the infrared radiation
passing through the doors, claims Bennett-Ireland.

The bulk of the heat is delivered by the
boiler/radiator system. A small flash
boiler mounted out of sight behind the
enclosure hood is connected via piping to
a radiator inside the hood. Heat rising
from the fire produces steam in the boiler.

Convecto-Pane Plus is made in four sizes to fit almost any fireplace, but it shouldn't be used on all-metal fireplaces, which could be damaged by the intense heat produced. Finishes? Your choice of satin black or antique brass.

## HEAT EXCHANGER

Simplicity is the word for the Hearth Aid heat exchanger: simply constructed, simple to install, and simple to use. Basically, Hearth Aid is a four-foot length

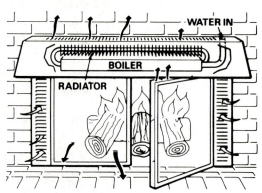

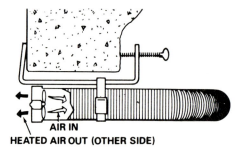

**Convecto-Pane Plus** features double-glazed enclosure, plus boiler and radiator hidden behind enclosure hood as shown in drawing. It uses convection to capture much of the fire's heat that normally escapes up the chimney.

The steam expands into the radiator. Air flows over the radiator, picking up heat, flowing out of the vented top of the hood into your home. As the steam cools, it condenses and flows back to the radiator to begin the cycle again. (If any water escapes through the air vent, you can easily replace it.)

The boiler/radiator system is said to deliver about 10,000 Btu per hour automatically, without blowers, so it keeps on working during power outages.

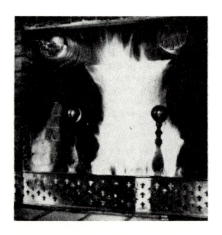

**Hearth Aid** produces the most heat when flames wrap around the heat-exchanger tube. Air is driven through the tube by a small 115-volt blower. Drawing at top shows how Hearth Aid clamps to fireplace lintel with a simple bracket.

of five-inch stainless-steel tubing that is C-shaped, with a small electric blower mounted in one end.

To install, you clamp the unit to your fireplace lintel—about two minutes' work. Not all fireplaces will take this type of installation, but Thermalite (Box 658, Brentwood, Tennessee 37027) gives clear instructions for alternate mounting. The sloping lintel of my fireplace wouldn't accept the clamps, so I had to drill two holes in the lintel, then attach a pair of brackets with sheet-metal screws. Installation time: about 10 minutes.

With the unit installed, you simply run the wire for the blower to the handiest outlet and plug it in. Build a fire and you're in business. As soon as the heat exchanger heats up, the blower turns on automatically to force air through the tubing. The motor has a variable-speed control—best left at full speed for peak efficiency.

Thermalite claims the Hearth Aid can produce over 20,000 Btu per hour. The blower costs about a nickel a day to operate.

You shouldn't use the system during a power outage, since excess heat may damage the unit. However, you can remove it in seconds and use the fireplace by itself if necessary.

The unit I tested had no glass doors to reduce heat loss, but Thermalite has since introduced Hearth Aid II, which does include a glass enclosure. This should improve performance dramatically.

# Return of the chimney sweep

*The soaring popularity of wood-burning has revived an ancient profession—you might want to try it yourself*

## By Al Lees

That fat old gent in the fur-trimmed cap isn't the only one hopping down chimneys these days. Most of the folks you see up on America's roofs each winter are young and slim—and sport headgear that's every bit as traditional: a black top hat.

They're members of a growing fraternity of professional chimney sweeps (nearly 2,000 in the U.S.) whose skills have become ever more essential as our nation's romance with the wood fire continues to heat up. Chimneys that haven't been used in years now regularly sport a plume of smoke—and rooflines that never sprouted anything but TV antennas now bristle with flue pipes as the sale of wood stoves continues to soar.

One consequence has been a dramatic increase in what used to be a very isolated problem—chimney fires. We have lost our familiarity with wood-burners; many novices don't realize that frequently used chimneys need regular cleaning.

When hot, volatile gases released by burning wood enter the cooler flue lining, they condense to form a thick, gummy substance called creosote. Fire after fire, the creosote continues to build and, if left unchecked for a few years, can turn a chimney into a potential volcano. It is possible to minimize such a build-up; but no matter how careful you are, the day will come when your chimney needs cleaning.

The National Fire Protection Association recommends that chimneys be inspected annually and cleaned, if necessary. Other sources say: "Clean after each cord burned," but if your woodburner is a tight, slow-burning stove, your creosote buildup could be even faster. Many local fire departments offer free inspection service, but how do you go about locating a sweep to do the cleaning?

In most communities, it's as easy as checking the classified's in the local paper. You might even keep your eye out for a black-clad figure in a top hat on neighboring roofs, though a better way to spot a sweep out on a job is to see his or her van at curbside as you drive through town. Most sweeps now work from the

hearth *up*, like George Ross in the photos on the following pages.

Ross noticed an ad for the August West Systems (one of the sources of gear and instruction listed at the end of this article) in *Popular Science* and thought his 19-year-old son might like to give chimney-sweeping a try. But the more Ross thought about the idea, the more he decided it might be "kind of a kick" for himself.

George lives in Richmond, California, and works for BART (Bay Area Rapid Transit). He keeps looking for productive free-time activity. With his three children grown, he felt the need for part-time work that would be a change of pace from his sedentary job. Why not chimney sweep-

ing? He mailed his check for the kit and hasn't regretted it.

"It's a hell of a lot different from my regular job," he says, "in that it gives me good exercise and the chance to meet new people. The pay's not bad either—$40 for a chimney I can do in an hour—and if BART goes out on strike like they're threatening, this sweep job'll come in real handy."

George's wife, Trudy, thinks the new business makes a good diversion for her husband: "I don't mind when he comes home dirty as long as he's smiling through the soot. He's always enjoyed clowning around with folks and his top hat now gives him a perfect excuse."

His job has hidden benefits for Trudy, as well. "It's good luck to kiss a chimney sweep, you know," she says with a wink.

Whether you want to invest in a new career, or just acquire the simple tools to tackle your own chimney, there are now a number of sources for quality gear. One of these sources has just come out with a special do-it-yourself kit.

I was persuaded to clean my own chimney when I read up on the frequency and ferocity of chimney fires. There were over 40,000 such fires in the U.S. last year, resulting in some $23 million damage—including the loss of complete homes. From all accounts, a chimney fire can be a terrifying experience, as the ignited creosote that lines the flue sucks up an ever-increasing draft, and flame and sparks roar out the chimney top.

How do you deal with such a fire? In response to a *Popular Science* article on the subject, many readers wrote in to ask—or to quarrel with the advice therein. Doug Benson of Port Charlotte, Florida, wrote: "I note you say to throw water into the stove (to quench the fire, I suppose). Back on the farm, in my youth, all heating was by wood, and many's the time I climbed on the roof with a bucket of cattle salt to quench a flaming chimney."

Actually, the article stated (in part): "If necessary, throw water into the stove and close the door. You may . . . warp the stove, but . . . the resulting steam should limit the intensity of the flue fire." But water is a last resort, whether for stove or fireplace (where it could crack hot mortar). The old ways are best, as this letter from R. E. Tatlock of Seymour, Indiana, proves: "Having grown up in the era when 'central heat' was a stove in the middle of the room, I remember grandma's method of quelling a flue fire. Close the draft and throw a handful of salt in the firebox."

## Here's how a California sweep

**After covering carpet** with dropcloth, George Ross brings in his vacuum and cleaning tools. He removes grate for wire-brushing, cleans ash from hearth.

**Removing damper** can be tricky since there are many types, but there's usually a cotter pin that releases it, giving Ross access to the flue above it.

**Fireplace itself is scrubbed** next with a wire brush. Scraper on back of this brush helps Ross remove any hardened deposits of creosote (note work light).

The vital thing is to close the draft caps (or damper) to cut off as much air supply as possible. (If you have coarse salt handy, by all means toss it in.) For a fireplace, wet an old blanket and tack it across the opening.

There are also special flare-type extinguishers for chimney fires that generate a mixture of gases to smother the fire. But the best policy is still prevention.

How about those powders that advertise: "Just sprinkle in your fireplace to burn away chimney soot"? All the sweeps I've spoken to say such products are of little or no practical use. Several told horror stories of seeing metal chimneys corroded by regular use of aluminum sulfate and other burn chemicals.

And what of those glassy tar-like deposits of creosote the flue-brush misses? If you're diligent, you may want to try burning these off with a high-heat torch. Sweeps tell me they encounter some old chimneys with deposits so anchored from years of refiring that the only recommendation they can make is to have the flue relined. But that's not likely to be your problem. Chances are, a thorough wire-brush scrubbing will remove all risk of a chimney fire for another year or two—or another cord.

## tackles a fireplace

**To clean flue,** Ross couples brush of proper size to flexible fiberglass rod. If flue is clogged with soot, Ross will start with smaller brush and work up.

**Quick-connect couplings** let Ross add extension rods to push brush through damper opening and up flue. With six five-foot rods, Ross can do two-story flue.

**Five paper bags** full of carbon and soot were removed from this very dirty chimney. His vacuum is to keep soot from drifting into room, not for cleaning.

## Wood stoves are a special problem

To learn about the cleaning of wood-stove flues, we collared a 26-year-old expert sweep from Connecticut at an August West regional workshop. Like most experienced sweeps, Steve Curtis prefers to work from the bottom of the flue as shown in the photo, though he'll climb onto a roof when the job requires (as he demonstrates in the silhouette photo, page 109). High-efficiency "air starvation" stoves create more creosote buildup than open fireplaces because of their slow burning rate and relatively cool flue gases. "A sealed overnight fire," says Curtis, "will deposit creosote even with dry hardwood logs. Always open the draft caps and let the fire burn hot for at least five minutes every morning, and again just before you close it down at bedtime."

The worst stove hookup is into a sealed-off fireplace: Smoke fed into a cold firebox will quickly condense its creosote on the walls and flue. Then, in the spring, many people pull out the stove in order to enjoy an open fire during the non-heating season. They're likely to get more than they bargained for when the hotter-burning fireplace ignites the stove's buildup in the firebox and flue. A better way to retrofit a fireplace with a wood stove, says Curtis, is to run the stove's smoke pipe all the way up into the existing flue. Even then, chimney cleaning is required. In the photo, Curtis dismantles a proper flue connection through a bricked-up fireplace. This will give him access to the flue, into which he'll insert the brush on a flexible rod (foreground of photo, which shows all gear ready for the job).

"Even where the smoke pipe rises directly from the stove," says Curtis, "it's best to detach it for cleaning. Where it's impractical to remove a long enough section, you have to insert your brushes from the top of the flue. If the pipe has a complex run, with several elbows, you may even have to attach your brush to a plumber's snake."

The stoves themselves are often harder to clean than an open fireplace, and with all the disassembling and re-assembling, the job is likely to take longer. Fee is about $40.

## You can do the job yourself if you buy the right brush

Shown on roof of my home, I am using proper brush diameter for this double-wall metal stack. Sizing is important. Wire brush with hard bristles should fit flue diameter almost exactly (whether round or—for masonry—rectangular). Softer bristles can have slight deflection as you tug brush into chimney; but don't risk an oversize brush that could become stuck in the flue.

To clean this chimney, I'll first remove the cap under my right hand. Bottom end of brush shaft is threaded to take a screw-on fitting called a "loop" ($1.25 from Worcester Brush), to which second rope is attached and fed down flue. Helper, below at hearth, grasps this rope and alternates tugs with me, on roof, to move brush slowly down flue in a scrubbing action. Old system of hanging a weight from brush (heavy chain or a bag of sand or stones) is now discouraged, especially where chimney has an angled offset as this one does, lower down. To do effective job, you'd have to use 15-to-30-lb. weight. This not only makes sweeping heavy work, it presents a distinct risk: Should weight detach itself during scrubbing—or should roof-rope break—plunging weight could damage flue lining (an even greater risk with clay tile or firebrick).

Choose a cool day for the job (below 40 degrees) so there will be positive draft up chimney to aid dust control. As for control at the hearth end, there's some controversy about using a household vacuum cleaner. Suppliers of professional gear advise using a powerful industrial vacuum; they claim the home type is not only ineffectual (fine dust will pass through its filter and back into room air), but can be damaged by corrosive soot. If your home has a built-in vacuum system, vented to outdoors, you might want to risk it. Otherwise, try your shop vac or rent an industrial model.

# Fireplace inserts— Open-fire charm, wood-stove efficiency

*Why brick up your energy-gobbling fireplace? Upgrade it instead, and save the flames*

## By Susan Renner-Smith

"An awful lot of people who won't stick a wood-burning stove in their living room will use a fireplace insert," says J. Grimes, president of the Phoenix Corporation (Asheville, North Carolina). "And as these inserts become more efficient, they become a viable alternative to running a stovepipe through a sealed-up fireplace," he adds.

Why install an insert in your fireplace? Though comforting, an open fire is notoriously inefficient. Not only does most of the heat fly up the chimney, but the fire also gulps room air—air you've already paid to heat.

Faced with a choice between the romantic wastefulness of an open fire (at best, 15 percent efficient) and the higher efficiency (55 percent) of a solid-fuel stove, many choose the stove. But more and more are choosing to compromise with an insert. Basically a double-walled stove shoved into your masonry fireplace, the insert is about 40 to 50 percent efficient. You get reasonable efficiency, a neat-looking installation that takes up no floor space, and, with many inserts, the ability to see the flames through airtight glass doors.

## BOOSTING COMBUSTION EFFICIENCY

Like good wood and coal stoves, inserts control combustion by feeding the fire the minimum amount of air needed to keep it burning efficiently. Most units have both adjustable combustion-air inlets and front-controlled dampers (you wire open the existing chimney damper). All have tight-fitting doors to keep room air from being sucked up the chimney. Unlike a fireplace, the inserts return heat to the room by both radiation and convection. The space between the insert's firebox and outer wall serves as a convection chamber that warms circulating room air

**Heat from this roaring fire** warms more than the hearth. The Phoenix House Collier (above) can heat 2,000 square feet for eight to 12 hours on a single load of coal, says maker. (Like some other manufacturers, Phoenix also offers a wood burner with an optional shaker grate for conversion to coal.) The insert, with the addition of bolt-on side and top panels, converts almost any size fireplace to a forced-air heater. It works on the same basic prin-

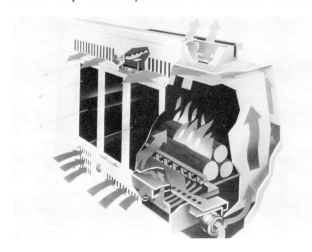

ciple as the Monarch Convert-a-Fireplace illustrated at left. Combustion air enters the center bottom vent. The firebox also has floor vents to admit outside combustion air. Meanwhile, cool room air flows into the convection chamber via side front vents. Side-mounted blowers speed the air's passage up and around the firebox; warmed air returns to the room via top vents. Factory insulation surrounds the insert, sealing it. Unlike many makers, Monarch (Beaver Dam, Wis.) markets the insert as a supplemental heater rather than as a furnace substitute.

(see diagram). Many inserts have optional blowers to keep air circulating. Some inserts, such as the Phoenix, also have baffles or fins inside the convection chamber to slow down the flowing air so it absorbs even more heat.

But how much heat do the inserts produce? "A well-run insert will produce 40,000 to 50,000 Btu an hour," says Dr. David Dyer of Alabama's Auburn University. Dyer has studied inserts and has used one in his home for three years. "It heats my 3,000-square-foot house," he reports, "if I feed it every two or three hours."

Dyer's point? Despite most manufacturers' claims, the inserts are not as effective for nighttime heating as a free-standing stove. "It gives off enough heat for this climate [Alabama], where winter night temperatures don't fall below 20-degrees," he reports, "but not enough for really cold climates."

Still, consumers battered by rising oil prices may welcome the chance to shut down the furnace during the day.

An insert costs no less than a good-quality airtight wood stove—prices start at about $500 and go to $1,000 for units that can burn either wood or coal. Choosing one isn't easy, though. One source-book alone (*Woodstove Directory*, $3.50 from Energy Communications Press, Box 4474, Manchester, NH 03108) lists many different units. Dyer says a good insert meets the following basic criteria:

- It's been tested by a recognized lab using UL standards.

- The firebox wall is ½-inch-thick steel or cast iron. Firebricks aren't essential, but they will protect the firebox from eventual corrosion.

- The doors and the insert itself are tightly sealed to keep out room air. Some manufacturers factory-insulate, others recommend jacketing the insert with fiberglass batts.

- The glass is a high-quality, heat-resistant glass.

Instead of glass, some inserts have heavy metal doors. You open the doors and snap on a fire screen when you want to gaze into the flames. "Solid metal *is* more efficient," says Dyer. "But I have glass. I like to see the fire all the time."

Many units come with tempered glass to resist breakage. Unfortunately, tempered glass shatters if it gets hotter than about 700° F, and a really hot fire can heat the doors above that. "It's better to get a heat-resistant glass like Corning's Pyroceram, or at least an intermediate type like tempered Pyrex," says Dyer. Several makers offer an insert with double-pane glass doors—heat resistant on the inside, impact resistant on the outside.

A final caution: Like any airtight stove, a fireplace insert should be used only with a perfect chimney, which should be cleaned regularly to prevent creosote buildup. But, advises Dyer, "if you run a wide-open, blazing fire for 30 minutes each time you start up, you'll vaporize all that creosote before it builds up."

# How to choose an energy-efficient fireplace

*A testing and labeling program makes it easier. But you have to know what the numbers mean*

## By Tom Rehberg

The warm glow of a fire crackling in a fireplace adds a special dimension to any room. But that charming source of warmth—both physical and psychic—has been attacked in recent years as an energy waster. and indeed, many fireplaces do waste energy, under some conditions.

The traditional fireplace warms the room by radiating heat from the fire to other surfaces: walls, furniture, people. It draws combustion air from the room—air that most likely was heated by your furnace at considerable cost—and wafts it up the chimney. Meanwhile, cold air from outside seeps in through cracks and openings in your house to replace that air. When outside temperatures are above 30° to 40° F, a typical radiant fireplace can add heat to your house. Below that, it probably is a losing proposition.

As the fire smolders and dies, the open radiant fireplace can be a heat loser even at warmer outside temperatures. The low fire produces very little heat, but warm room air continues to zoom up the chimney—perhaps all night unless someone stays up to close the damper after the fire dies.

A fireplace doesn't have to be a wastrel, however. Just about all makers of manufactured fireplaces now include energy-efficient models in their lineups. These are heat-circulating fireplaces with double-wall steel fireboxes that act as heat exchangers. They transfer heat by convection as well as by radiation. Some also have glass doors to isolate the fire from heated room air and ducts to bring in combustion air from outside.

Energy-efficient fireplaces can contribute substantial amounts of heat to your house, supplementing your furnace and perhaps reducing your total fuel bill. They are available as zero-clearance units, which you can install directly on wood floors and against framing studs. This type needs no masonry surround or foundation, so it's easier and cheaper to install than a conventional fireplace.

Until recently, however, shopping for an efficient fireplace was a confusing matter. Manufacturers made widely varied efficiency claims and substan-

117

tiated those claims with different tests—or none at all. Now the Wood Heating Alliance (a trade association of manufacturers, distributors, and dealers) is sponsoring a voluntary testing-and-labeling program. Fireplaces submitted under the program are put through uniform tests at Auburn University, Auburn, Alabama. Results are recorded on labels, which are placed on the fireplaces (see sample).

The tests evaluate each fireplace at three firing rates: high, medium, and low. The amount of heat transferred to the room (in Btu/hr.) is measured, and the efficiency is calculated by dividing the energy flow into the room by the total energy content of the wood. The efficiency is shown as a percentage. Finally, the volume of room air required for combustion is computed and recorded on the label.

These labels let you fairly compare different fireplaces from different manufacturers. But unless you understand the difference between gross heat and net heat, they could be misleading. Gross heat counts all the heat delivered to the room by the fireplace, but it does not subtract the heat lost when warm air is drawn out to support combustion. Net heat, on the other hand, does consider the energy used to warm the cooler air that filters into the house to replace escaped room air.

In the calorimeter room at the Auburn laboratory where the fireplaces are tested, the temperature is maintained at 70 to 75 degrees F. So is the air supplied to the room. Thus the heat output (Btu/hr.) that the tests measure is net heat only when outside temperatures are 70 to 75 degrees—not exactly fireplace weather. At any other outside temperature, it is a measure of gross heat.

Why doesn't the label give net heat output over a reasonable range of outside temperatures? For one thing, because the whole subject of outside combustion air is a raging controversy within the industry.

While many manufacturers and theoreticians see it as a way to boost fireplace efficiency, others aren't convinced. "What happens to combustion efficiency when the air being fed to the fire is 20 below zero?" ask the skeptics. "And how much does that cold incoming air cool off the heat exchanger?" No one can answer those questions with certainty yet, but tests to do so are underway at Auburn.

Meanwhile, the Wood Heating Alliance tests and labels are a compromise between the two sides. Gross, not net, heat is stated on the label. But another measurement, the amount of room air required for combustion (in cubic feet per minute), is also taken and recorded. A closed fireplace with an outside air source will take much less room air than an open fireplace without ducts from outside (though no fireplace is airtight). The relative importance of the room air consumed, however, is left for you to decide.

If, indeed, outside combustion air is an efficiency booster, fireplaces with this feature will probably perform even better (compared with their open-faced rivals) than the Btu/hr. and efficiency ratings on the labels would suggest. And they'll do so at an increasing rate as outside temperatures drop.

## SHOPPING FOR EFFICIENCY

The Auburn tests indicate that the efficiency of conventional radiant fireplaces ranges from −10 to 10 percent. Heat-circulating fireplaces with glass doors that use outside air sources typically range from 10 to 35 percent. To understand why the efficiencies vary so much, take a closer look at how heat-circulating fireplaces work. Near the floor, these units have ducts through which room air is drawn into an air chamber between the double walls of the fire-

**Heat-circulating fireplaces,** such as the Energizers from Superior shown here, heat by convection as well as by radiation. A double-wall firebox acts as a heat exchanger. The air passage must be totally isolated from the firebox, or the room will fill with smoke. The Energizer also uses glass doors to isolate the firebox from the room, and gets its combustion air from outside.

box. The air passes over the hot steel walls, rises, and exits to the room through high ducts. Some models use fans or blowers; others depend on natural convection.

One important design factor affecting the efficiency is the total heat-transfer area. Some have the doublewall air chamber on only three sides of the firebox; others surround all five sides (two side walls, back, top, and bottom). Given the same size firebox, the one with airflow over more surfaces will transfer more heat to the air. Furthermore, some manufacturers include baffles within the heat exchanger to provide a longer path for the air to travel. This means the air that's returned to the room will be hotter. When shopping, find out how many sides of the heat exchanger and how much, if any, baffling is included. If the manufacturer's literature doesn't tell you, ask your dealer.

Speeding up airflow through the heat exchanger can produce a greater *volume* of heated air, increasing fireplace efficiency in another way. Fireplaces that use blowers or fans on the air intake tend to be more efficient than those that depend only on convection.

## FIREPLACE TYPES AND SIZES

If you select a fireplace that uses outside combustion air, be sure that the doors form a reasonably good seal around the fireplace opening. If they don't fit well, combustion air will still be sucked from the room rather than from outside. Some fireplaces come with glass doors that are spaced out an inch or two from the fireplace. These doors may be decorative and serve as spark arresters, but they do not keep your heated room air from going up the chimney. Furthermore, installing new glass doors that fit snugly against such a

fireplace could create a safety hazard. Since each fireplace is engineered differently, your safest course is to buy only the doors specified by the manufacturer. And they should be UL listed.

A fireplace that uses outside air for combustion should have a convenient and effective mechanism to regulate the volume of air so that the combustion rate can be controlled. Multiple dampers on intake ducts are superior to one. They create insulating dead-air spaces when a fire is not burning, and reduce condensation and frost problems. Also, a minimum area of the firebox should be exposed to cold incoming air.

Some heat-circulating fireplaces are made to heat only the room they are in; others can be ducted to other rooms as well, but a little efficiency will be lost. To minimize heat losses, the duct length should be kept as short as possible and sharp bends avoided. Wrapping the ducts with noncombustible insulation will also help (check manufacturer's instructions). A fireplace shouldn't connect with your furnace ducting system unless it is specifically made for that purpose and instructions detail exactly how to do it. Local codes must be consulted.

A fireplace that's too large for the space you want to heat will perform less efficiently than one properly sized. A fireplace that's too small just won't do the job. One way to get a rough idea of how large your fireplace should be is to extrapolate from the Btu/hour output rating of your furnace. Divide the room area you want to heat by the total area of the house (or as much as the furnace was intended to heat), and multiply the answer by the Btu rating of your furnace. The result will give you a very rough estimate of the gross Btu/hour rating the fireplace needs to have. But keep in mind that many furnaces, particularly older ones, are much larger than they should be for the size of the house. Also remember that the

**How to read the label:** (A) Manufacturer and model number. (B) Test number: Each fireplace is tested at three firing rates—test 1 is low fire; test 2, medium fire; and test 3, high fire. (C) Wood lbs./hr.: pounds of wood burned per hour during each of the three tests. Figures tell you how much wood must be burned to achieve a particular heat output (Btu/hr., next column). If you know your heating needs, you can estimate your wood costs (a cord of wood weighs about 4000 pounds). (D) Btu/hr. output: heat output into the room at the three firing rates. (E) Percent efficiency range: determined by dividing the energy output into the room by the energy in the wood. (F) CFM room air required: volume (in cu. ft./min.) of room air consumed. The more room air consumed, the less net heat the fireplace will produce. The problem becomes more critical as outside temperatures fall.

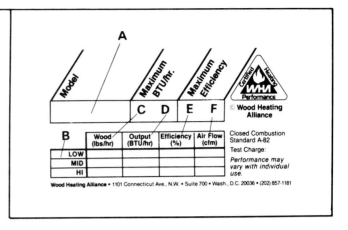

gross Btu rating of the fireplace does not consider any heat lost from room air used for combustion.

The physical size of the fireplace, as well as its Btu output, may be important to you. The size of a fireplace is given in terms of its viewing area, not the actual size of the firebox. Firebox sizes vary greatly.

## INSTALLATION CONSIDERATIONS

Zero-clearance designs make installing a fireplace much more manageable for the average do-it-yourselfer. But not all models are equally easy to install. Generally, one that heats only one room will be easier than one that is ducted to other rooms. Installation ease also varies from one make to another, so review the instructions before making your decision.

A fireplace with a low profile (the distance between its floor and the point where the chimney attaches) will give you more flexibility in routing the chimney. That can be critical in a room with a low ceiling. Some units take up so much vertical space that you can't make the necessary turns to get the flue pipe out of a seven-foot basement. Low-profile units give you more maneuvering space and let you put the opening higher up in the wall, should you prefer a raised hearth or firewood storage beneath.

Before you make your fireplace purchase, it's wise to know what accessories and replacement parts are available. Glass doors are fragile and will occasionally break. Blowers and grates may need replacement.

Beware of buying bits and pieces to make up a total fireplace package. Manufactured fireplaces are carefully engineered systems designed to operate at peak efficiency with the intended components. By taking a firebox from one maker, glass doors from another, and blowers from a third, you not only compromise the energy efficiency of the unit; you may also rig yourself a safety hazard. Be sure the entire unit has safety accreditation, such as UL listing. Review and compare warranties.

Finally, consider your ultimate plans for the fireplace. Might you one day want to burn gas logs instead of wood? Do you plan to add glass doors, outside combustion-air ducts, and heating ducts to other rooms at some later date? If you initially install a self-contained radiant fireplace, it may not be possible to upgrade in the future. Some accessories are available for some fireplaces, but you can't convert a conventional radiant fireplace to a heat-circulating model. If that's what you want in the long run, buying a cheaper model today could prove economically unwise tomorrow.

# Retrofit fireplace matches wood-stove efficiency

---

*Now you can have both heat and ambience.*

---

## By Dan Ruby

Wood-burning appliances can serve two functions: They can heat a room or house, or they can provide a crackling, colorful ambience. Until recently, they could not do both at once.

Traditional masonry fireplaces, long favored as romantic scene-setters, actually contribute little heat; in fact, they often steal energy from a room by consuming great quantities of already heated air. Wood stoves offer heat—up to 60 percent of the Btu potential of the wood, even more for catalytic types—but for many people they come up short on atmosphere.

That all began to change several years ago when new, prebuilt fireplaces were introduced. These use various arrangements of special ducts, doors, dampers, and outside air supply to raise efficiency to about 35 percent while retaining the charm of a brick fireplace. Now a new

manufactured fireplace is available that boosts efficiency even more—almost to the level of the best wood stoves, claims its maker, El Fuego Corporation.

Furthermore, El Fuego IV Zero Clearance Fireplace B-CF can be installed against any wall and on any floor (zero clearance refers to its UL listing that allows installation against combustible surfaces). This permits some imaginative placements: recessed into a wall, placed in the interior of a room to double as a divider, or angled at 45 degrees into a corner. For the average do-it-yourselfer, installation is a weekend job. List price of the unit is around $1,000.

The company's efficiency claim is backed by the rating of the Wood Heating Alliance, an industry association established to devise wood-burner standards. In tests at Auburn University, the Alliance found that El Fuego IV produced 40,000 Btu per hour while burning as little as eight pounds of logs—for an efficiency of 57 percent. In comparison, an average masonry fireplace uses three times as much wood and produces half as much heat.

The tests showed that El Fuego's high performance resulted from the very low consumption of room air necessary to

**Zero clearance** means no difficult masonry work. Simply frame an enclosure with 2x4's and install your choice of facings: brick, stone, paneling, tile, or painted wood. Ducting can be routed to other rooms to distribute heat where needed.

feed the fire: only 13 cubic feet per minute. The tightest of wood stoves consume about 10 cfm of air; a masonry fireplace uses 400 cfm.

The fireplace operates on natural convection with no need for electric fans or blowers. Room air is drawn in at the bottom of the unit, circulated around the fire-box, and ducted back into the room at two levels above the mantel. Tempered-glass doors over the fireplace opening leave the fire visible while allowing only a tiny flow of room air to the fire. The fire burns with the damper 92 percent closed to allow smoke but little heated air to escape up the chimney.

## Flue Flusher

Push Flex Klean up your chimney and creosote comes tumbling down, says Plastic Techniques of Pennsylvania, Inc. (Box 1449, Scranton, Pa. 18503). The flexible plastic rod comes in five-ft. lengths; the disc, in two shapes, many sizes.

## Spark Blocker

If your chimney liner is cracked, even one stray spark can cause a fire. Renew-a-flue makes old masonry chimneys safe, even when used with high-temperature fireplace inserts, says Duravent (Box 2249, Redwood City, Calif. 94064). The stainless-steel chimney-liner kits start at about $295.

# 5 WOOD STOVES

Wood stoves are fashionable again, and with good reason—for many of us fortunate enough to live in a wooded area or having access to a woodlot, they provide heat for our homes from a free energy source. In reasonable quantities, gathering firewood can even be fun and good exercise.

Wood stoves are efficient, too—at least when compared to a fireplace. New catalytic models even surpass the efficiency of central heating plants.

But there are two drawbacks. Most wood stoves today are of the "airtight" design, using draft caps to throttle the supply of combustion air to an otherwise sealed firebox. This lengthens the burning time, but the smoldering effect results in low chimney temperatures and the accumulation of a tarry precipitate called *creosote,* which can lead to dangerous chimney fires. Also, the heat generated is all too often confined to one area of the home, an effect you've probably noticed where the room containing the stove is overheated while the remainder of the home is cold.

You can change all that, and we promise that when you do, you can enjoy your wood stove like never before—in maximum comfort and safety.

# Wood-stove ducting

*To spread the comfort, add or extend ductwork*

## By Evan Powell

Owners of wood stoves I've talked with, particularly those using air-starvation types, are pleased with the basic performance, but they want more comfort. They're baked in the overheated room where the stove is located, and shiver in other rooms.

The problem is that a wood stove is a point source of heat. Its ability to circulate air is limited by walls, partitions, and furniture. In a large open area, a wood stove probably could heat the 3500 square feet claimed by sales people. But houses just aren't built that way. Another problem is that stoves put out "raw" hot air, not the filtered, humidified air we're accustomed to.

I tackled these problems in two test installations that got a lot of use last winter. My conclusion is that a few simple, low-cost modifications to existing warm-air ducts in a home can integrate a wood stove into the heating system. The result is a better distribution of air for greater comfort. In most cases, these modifications can be a do-it-yourself job.

## MOVING HEAT

One element of every warm-air furnace is a supply plenum. It traps hot air flowing from the furnace and channels it into individual ducts. Normally, room air is picked back up through an intake near the floor level, and returned to the furnace where it is reheated and recycled.

When a wood stove is in operation, the room it occupies is similar to a plenum, since it's collecting most of the heat from the stove. Removing some of that air and distributing it to other areas of the house serves a dual purpose: It *raises* the temperature in distant rooms and *lowers* the temperature in the room where the stove is located to a comfortable level. You don't need (or want) a direct connection to the stove—just utilize the room as the heat storage area it is.

Simply running the furnace fan does this to a limited degree, too, but usually the return-air grille opening is poorly positioned. It doesn't pull much heat back into the system. When I located the return-air grille near the ceiling, I found that it made a substantial difference in comfort levels throughout the house.

Our test installations involved two very different problems. On Chestnut Mountain, a free-standing stove is situated in a two-story "great room." A loft and balcony office open into this area upstairs. Using the stove alone, the results were predictable: When the lower area was

129

comfortable, the upper area was too hot. And the basement was frigid.

One extension of the main return-air duct to the upper level allowed me to add a return-air grille in both the loft and balcony areas, near the peak of the cathedral ceiling. Now I just flip on the blower and within minutes all areas are brought to a more comfortable temperature.

At the second installation, the fireplace is located upstairs in the upper-level living room with the kitchen and master bedroom. The den and three other bedrooms are downstairs. A fireplace insert with a blower was mounted in the fireplace. The duct from an existing low-return intake was extended to near the ceiling level. Without the blower running, it was often necessary to heat the stove room to 80 degrees or more to maintain comfortable temperatures in other rooms.

With the redistribution blower running, temperatures in the living room dropped to a much more comfortable level. Even during 15-degree nights, the lowest temperature recorded in a downstairs bedroom was 58 degrees.

## MODIFYING DUCTS

To redistribute wood-stove heat in your home, analyze your problem and the existing return-air duct system. If a return intake grille is not located near the stove, you can run a new duct and connect it to the return-air system below the floor level. Otherwise, an extension should suffice. Often, you can do this by removing a sheet of paneling, or extending the return-air duct from behind if it runs through a closet nearby.

You can use any approved duct material, such as sheet metal, for the extension. I used Johns-Manville Duct Board, a heavy foil-backed, rigid fiberglass material that can be formed with simple tools into rectangular ducts. It's available from both insulation and heating suppliers in 4-by-10-foot sheets. Duct Board makes do-it-yourself jobs easier; it's quiet, self-insulating, and has a built-in vapor barrier.

In both test installations, supply ducts were already insulated. That's a must for low-level heat distribution if the ducts are located in an unheated area.

When a low-to-high-level duct modification is made, a damper can be installed to select between the two intakes. A high wall return is the most efficient place to pick up return air for central air conditioning as well as heat redistribution.

The value of winter humidification in homes is generally well accepted. Humidity levels around 35 to 40 percent eliminate problems of furniture and wood drying, and make you feel more comfortable by slowing the evaporation rate of skin moisture. You can be perfectly comfortable in a room of 65 to 68 degrees with an acceptable humidity level, and feel cold in a dry room at 80 degrees.

Don't forget that relative humidity (RH) is relative. Cold outdoor air may be at 90 percent relative humidity. But when it enters your house and is heated to 68 degrees, its relative humidity can drop below 10 percent. And that's drier than the Sahara Desert. It doesn't matter how fast or slowly it's heated or what the heat source is. It's the change in temperature that makes the difference in the humidity level.

But humidifying can be a tough job. The Aqua-Mist Humidifier people offer an answer: an unusual combination of spray and evaporative humidifier principles. It's available in a bypass unit—the kind I used—or in models that fit under or on the side of the main warm-air supply duct. The installation worked (see photos, page 133), maintaining reasonably good levels of moisture.

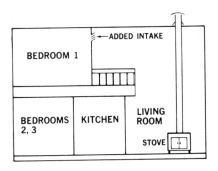

**Free-standing stove** is located in a two-story "great room." In this installation modified by author, rooms in the upper level overheated while those below were chilly. Intake duct for redistribution of heat was extended to reach near peak of ceiling. Johnson Control filters were fitted through grilles (left).

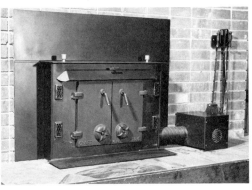

**Fireplace insert** (above, left) was added to fireplace located in large upper-level room of second upgraded installation. Initial setup provided little heat to rooms on lower level. An extension from lower grille provided high-level intake for redistribution. A manual damper isolates either grille as desired.

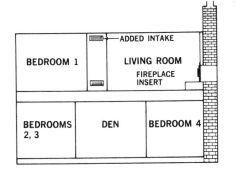

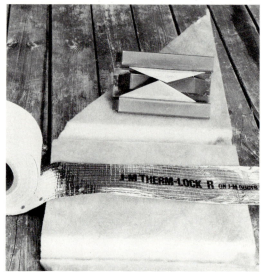

**Notching knife** is pulled through Johns-Manville Duct Board (top, left). Special V-notching tool (top, right) can also be used for DIY ducts; heat-sensitive tape seals corners and joints. Board is folded (above, left) to form duct corners. Taping (above, right) completes job for rigid installation free of rattles and vibrations.

**Bypass humidifier** can be installed between supply and return-air systems. Adjustable humidistat controls moisture and comfort level of air within home.

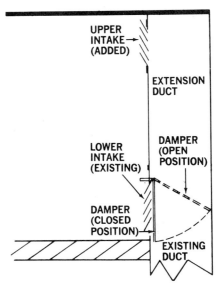

**Return-air-duct extension** should reach as close to ceiling as possible. Top intake grille is installed so louvers are directed upward. Manual damper allows lower grille to be used with conventional warm-air heating system. The damper is closed for high-level intake during redistribution or during air conditioning.

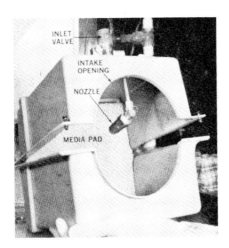

**Aqua-Mist humidifier** is especially designed for low-temperature applications. Nozzle sprays atomized mist against media pad positioned directly in path of air flow. Pressure differential between supply and return-air systems causes air flow through humidifier. Other models are beneath or on side of ducts.

## ADJUSTING TEMPERATURES

Homeowners often complain that the thermostat is located in the same room as the fireplace. With a heat-redistribution system, that's desirable because it gives you a manually integrated "hybrid" system. Just set your furnace thermostat to the lowest level at which you wish to maintain temperatures. If you're away from home often, you may want to consider a replacement thermostat that can be set as low as 40 degrees. When the wood stove is used, the higher temperatures will override the thermostat and the furnace won't run. If you're away from home or the fire dies down on a cold night, the furnace will run as often as necessary.

At Chestnut Mountain, the marriage between the wood stove and our Coleman high-efficiency heat pump proved to be practically ideal. A heat pump is very efficient at temperatures down to 40 degrees F or so. I like to crank up the wood stove on the first really cool day, but it's not very efficient in mild temperatures since you must keep the air shutters nearly closed. Allowing the heat pump to maintain temperatures during mild weather and using the wood stove when temperatures get in the 30's and below utilizes both heat sources under the conditions at which they operate best.

You don't have to start with a heat pump to reap the rewards of saving energy and money while enjoying the atmosphere and comfort of your wood stove—effectively. Any warm air heating system already has the means of distributing heat throughout your home, and there's almost always some way to utilize it effectively.

Your wood stove is a substantial investment. A few more dollars and a little time can integrate that heat source into your home's present heating system, and allow it to operate more efficiently and deliver filtered, humidified air as a bonus.

# How to control creosote for safer woodburning

*New scientific research shows that much of what we "knew" was wrong—and even dangerous*

## By Evan Powell

Heating with wood is satisfying in a lot of ways. But it has its dangers, too. The main one is creosote buildup. Creosote is a mixture of tars and organic particulates from wood combustion that condenses in droplets of moisture that form on the inner liner of a flue or chimney. If this buildup becomes severe, it can ignite and burn fiercely.

A creosote chimney fire can burn at temperatures approaching 2,000 degrees F—600 degrees hotter than a brazing torch. In a defective chimney, or in one damaged by a previous fire, it can cause severe damage and even set your house on fire. While a fireplace can generate some creosote, the problem is far more serious with wood stoves, especially air-starvation types.

What can you do to prevent creosote buildup and its attendant hazards? A lot has been written on this subject, but now there's important new information that changes many of the principles we've been taking for granted. It grows primarily out of work done at the Woodburning Laboratory at Auburn University, perhaps the country's leading laboratory devoted to finding ways to burn wood more efficiently and safely.

Dr. Timothy Maxwell, I learned on a recent visit, has a way of getting your attention by expressing some conclusions that are completely opposite to principles that have long guided wood-stove users. Some examples:

- *The amount of creosote buildup from a wood stove is not dependent on the type or dryness of wood.* Conventional wisdom is that pitch pine and similar woods generate far more creosote than, for example, dry, seasoned hickory. Not so, says Maxwell; in fact, the hickory may yield slightly more creosote.

- *Secondary air input (introducing room air to the volatile gases just above the flame in an "airtight" wood heater) doesn't reduce creosote buildup; it may even hurt.*

- *Proper sizing of stove pipes and chimney materials may be more important than the types that are used.*

These findings can be of tremendous importance to you if you plan to get all or

135

**Spark arrestors** prevent stray particles from reaching roof, but also collect a coating of flammable creosote; check regularly. Field glasses provide a quick close-up look. Don't mix fittings or accessories from different chimney brands.

part of your heat from a wood stove. They also demonstrate an important principle. As Dr. Tom Pruitt, manager of the lab, told me, "Burning wood is the oldest form of controlled heat known to man, but we know far less about it than any other heat source."

Many homeowners today fall into one of two categories at extreme ends of the spectrum—they completely disregard the creosote hazard, or they are so frightened by what they have read and heard about chimney fires that they won't use any form of wood heating. Either case is unfortunate.

Dr. Maxwell's work may help solve the problems. He doesn't claim to have all the answers—his conclusions are based on initial experiments—but after applying elements of these theories to observations I have made, and questioning other experts in the wood-heating industry, my initial skepticism has been replaced with the realization that if he's not entirely right, he's at least on the right track.

## HARVESTING CREOSOTE

In his lab, Maxwell has a dramatic way of proving his points. There, creosote precipitate is collected from a unique flue passageway that is studded with thermocouples and surrounded by three water jackets through which cold water can be passed in sufficient quantity to provide any desired flue-liner temperature. At the base of each of these three sections of the flue is an inverted cone that forms a trough. Precipitate flows into these troughs, then is drained through a pipe into graduates. After collection, a sample of the precipitate is placed in a centrifuge to separate the "pure" creosote from moisture and other impurities. Using this method, Maxwell discovered that there is little deviation (only 5 to 10 percent) in creosote yield from different types and drynesses of wood, though the volume of total precipitate (including water) is much greater for green wood.

**Test installation** at Auburn University's Woodburning Laboratory has conventional stove with elaborate exhaust system. Octopus-like maze of tubing carries water and steam into jacketed flue to control the temperature of its liner.

**Creosote/water mixture** flows from pipe (left) in test setup. Collected fluid is put in centrifuge, then analyzed to determine how much creosote (vial in inset) has been produced. In chimneys, creosote may be hard or soft and tarry.

On my desk is a vial of this vile substance, a centrifuged sample from the Auburn lab. The tarry liquid is encapsulated with a small amount of liquid precipitate to prevent it from hardening and drying out completely. It's mean and active, looking much like a brown and black "lava lamp." Over an hour ago, I inverted it; still, every few minutes a funnel of the stratified black layer at the top dips down like a miniature slow-motion tornado.

Where does all the liquid come from? Not just from the moisture in wood, as I learned when we conducted our own experiments. We used single-wall pipe outdoors in cold weather to lower flue temperatures purposely so creosote would be deposited, and found that it built up (and fast) regardless of the type of wood we used.

In the Auburn tests I saw recently, researchers put firebrands in a Kickapoo stove. These "brands" are a standard fuel used for such testing, consisting of ¾-by-¾-inch strips of dry Douglas fir that have been baked to remove all moisture. To my surprise, five minutes after the brands were ignited, fluid was pouring from the collection pipes like water from a faucet. How is it possible? Dry Douglas fir contains six-percent hydrogen by weight. As the fir burns, one pound of hydrogen combines with oxygen to yield nine pounds of water. For every pound of dry wood you burn, you create a half pound of water.

Even the "green pine" theory has begun to sound logical. In an open fireplace, the added moisture content of green wood could indeed lower flue temperatures to a point where condensation would occur more readily. A problem could exist that might not develop if dry hardwoods were burned.

But in an air-starvation type of heater, where creosote problems are worse, the damper is usually adjusted to provide the desired heat output that maintains room temperature. The damper becomes an "equalizer," since you tend to provide more air for green wood than for dry wood. Under these conditions, flame and flue temperatures would be about the same for both woods.

## SECONDARY AIR

This principle is touted by most manufacturers and sales people for "airtight" wood stoves (a real misnomer). It's supposedly the best way of extracting the last bit of heat from gases—and eliminating creosote.

But there are two things wrong with this theory. First, a wood fire is not constant; it doesn't give a steady, controlled output as does an oil or gas burner. Second, provision for secondary air on most air-starvation heaters is fairly crude. It's usually a small, unsealed opening or crack near the top of the door on the stove.

According to Maxwell, this is effective (if at all) at only one point in the flame cycle: as it rises and falls when wood is added. At all other times, it may actually *promote* creosote buildup by lowering flue-gas temperatures.

This may be one area for future improvement in these heaters, since secondary air does seem to work in more sophisticated applications, such as central wood-burning furnaces.

Also, according to Maxwell, the race to extract the last bit of efficiency from an air-starvation-type heater contributes to creosote buildup. This efficiency is gained by adjusting a simple damper that regulates the input of primary air—the air admitted directly to the fire for combustion. Closing this damper lets wood burn a long time, with a corresponding drop in temperature. At the lowest levels, only a cool, smoldering fire is present.

Maxwell feels that opening the damper slightly or, better yet, building in a stop to prevent complete closure would greatly improve the situation. Efficiency should drop only slightly. Other experts generally agree. One told me he could cure the creosote problem by drilling a hole through the damper. Others suggested that wood heating is not intended for mild weather. If it is too warm in the room for a moderate burn, they say weather conditions aren't cold enough for wood heat.

## METAL CHIMNEYS

Much of the controversy about creosote and wood heating centers around the type of chimney system used. There are three basic types (see diagrams) of "prefab" chimneys sold for solid-fuel heating equipment: double-wall solid-pack insulated; triple-wall modified thermosiphoning; and true thermosiphoning.

Based on the recommendations of many manufacturers, only solid-pack insulated or triple-wall modified chimney installations for air-starvation heaters—not the true thermosiphon types—should be used. But you should understand chimneys to determine the safest system for your home.

While it's difficult to pin them down, most experts I interviewed say they prefer solid-pack insulated metal for their homes. Triple-wall modified and masonry chimneys tie for second place. The reason for these choices is simple. The inside flue liner of insulated and modified chimney types runs hottest. Air-starvation equipment typically produces low flue temperatures, so the hotter you can keep that inside flue liner, the better.

The air wash in true thermosiphoning chimneys keeps the inner liner so cool that precipitation is much more likely to occur. But if a chimney fire does occur, thermosiphoning pipe remains much cooler than other exhausts. Ray Hemmert of Majestic believes the advantages of thermosiphoning chimneys make them the right choice for his company's fireplaces. But he *doesn't* recommend thermosiphoning pipe for airtight heaters.

Hemmert says that problems with thermosiphoning chimneys arise when the five- or six-inch outlets on smaller air-

## How metal chimney systems compare

**Solid-pack insulated chimney** has insulation filling cavity between liner and outer shell. Interior surface of liner generally runs hotter than other chimney types. True triple-wall thermosiphon type uses convection currents to cool inner liner with air wash. As air surrounding liner heats, it rises and flows out top. Cool outdoor air rushes in outer cavity to replace it, cooling outer shell. Action varies with flue-temperature changes. These systems have worked well with fire-

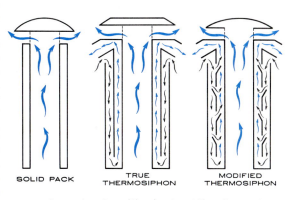

SOLID PACK          TRUE THERMOSIPHON          MODIFIED THERMOSIPHON

places, but supercooled liner can generate creosote with air-starvation-type stoves. Modified triple-wall thermosiphon has baffles that limit air wash to maintain hot inner liner with cool envelope. Experts interviewed by author preferred the solid-pack design.

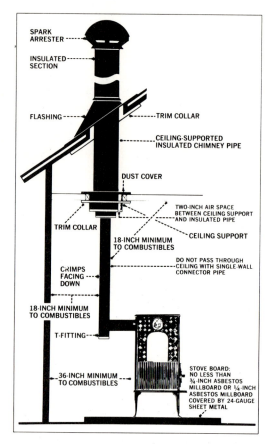

**Recommended installation** for Jøtul stove shows clearances and components to minimize hazards. Note the cleanout T-fitting in smoke pipe near stove. Chimneys should be kept indoors. Use insulated pipe outside heated areas. (Courtesy Kristia Assoc.)

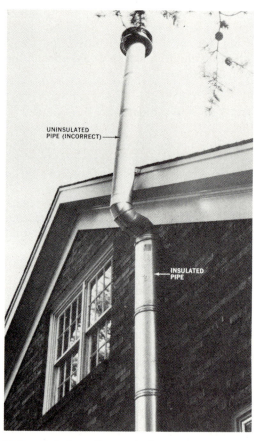

**Uninsulated sections of pipe** at top of chimney cause creosote precipitation that builds up rapidly in area below. Installations like this are all too common. If a chimney fire does occur, uninsulated sections would then offer little protection.

starvation heaters are adapted to a larger thermosiphoning pipe. The large liner area keeps temperatures much lower than they would be if the vent were sized properly. Hemmert maintains that true thermosiphoning pipe should pose no problem on large wood-burning equipment for which eight-inch or larger pipe is recommended.

But the state of Maine no longer approves the installation of prefab metal chimneys in conjunction with wood or combination wood/oil furnaces. The Maine Department of Public Safety has issued a news release warning homeowners about chimney fires, and State Fire Marshall Donald Bissett tells me he's considering extending the ban to cover air-starvation woodburning stoves if the Energy Testing Laboratory of Maine can provide documented evidence that such chimneys present a hazard.

This private facility is operated by Carl R. Flink, department head of heating and air conditioning at the Southern Maine Vocational Technical Institute. But I could never reach Flink there, despite repeated calls. No Maine official I spoke with could give me any details about the methods or scope of this lab's tests.

Just because a wood stove is connected to a masonry chimney is no cause to rest easy. Many masonry chimneys would have a difficult time meeting the requirements for prefabs, according to UL's Charles Gibbons, who also feels that much of the present data on chimney fires leaves more questions than answers. At UL's Northbrook, Illinois, facilities, a research project is underway in which chimney fires will be induced, and all types of prefab chimneys tested.

## COMBATING CREOSOTE

You'll never entirely prevent creosote, but you can reduce dangerous buildups. Despite some confusion, we know from our present experience that the following practices will virtually guarantee you'll never have a chimney fire. These rules apply to both existing and new installations.

1. *Use UL-listed wood-burning equipment and chimney material. And be sure that the unit is installed to the manufacturer's instructions.*

   The smoke pipe and chimney should be secure and supported, and should be the size recommended for the equipment. Also, each section should be installed with the crimped end *down* so that precipitate runs back into the heater. Most manufacturers supply adapters for this purpose.

2. *Inspect the flue and chimney regularly to check for a tarry buildup of creosote.* The easiest way to do this is to install a T-fitting with a removable cap in the smoke pipe. Underwriters recommends that inspection be twice a month, and some experts say weekly or even semiweekly. Under extreme conditions, I've seen precipitate form in less than a day. If there's a severe buildup, don't use the equipment until it has been cleaned. If a chimney fire occurs once, it will probably happen again. Find the cause of the fire and *have the chimney inspected for damage* before putting it back into use.

3. *Know your clearances.* Those specified in your equipment instructions are *minimum* clearances; it's better to have still more distance. Remember that combustibles such as furniture or carpeting should never be placed within the minimum-clearance boundaries.

   Les Githins of General Products Company recommends a "touch test" for every installation. With the heater in full operation, feel the wall and surrounding areas. If hot, you can protect it with millboard or a metal shield spaced one inch from the wall.

4. *Starting with a clean chimney, at least once a day, and always before adding fresh fuel to the fire, open the damper and let the stove burn hot for 15 minutes or so.* This practice burns away small amounts of creosote at a much lower temperature than is required to remove thicker coatings that have built up over a longer period of time.

   The value of this practice may account in part for statements such as that of Yukon Industry's David Tjosvold. He told me "we have never encountered a buildup of creosote except in a few rare instances where insufficient combustion air was available." Yukon makes central wood-fueled furnaces that by design fully open a damper when the thermostat calls for heat. Safety is built in.

5. *Do not use chemical cleaners.* I have experimented with several of them during the past year. Those containing copper sulfate seem to work reasonably well. According to the cleaner manufacturers, the copper will coat any soot in the chimney, acting as a catalyst to allow the soot to burn away at lower-than-normal temperatures. But experts feel this could reduce chimney life, and it may also give you a false sense of security.

"Occasional use probably doesn't hurt anything," Hemmert says, "but users typically don't follow instructions for frequency or amount of use. The sulfur derivatives and that much moisture combine to form sulfurous or sulfuric acid. Over a period of time this can cause damage that goes undetected. Since metal chimneys are typically designed for a lifetime of 35 to 40 years, there's no way of knowing how much damage such chemicals cause."

Learning to harness the energy potential of wood is much like working with electricity: It's respect, not fear, that allows us to make full and safe use of it.

# Stove-centered house designed for efficient wood heating

*The look is dramatic, but the system is simple—and the house is prebuilt*

## By Evan Powell

On a cold, rainy day in February, I arrived in Gainesville, Georgia, to check out the new Støvehaus. This innovative house, designed by John Odegaard, vice-president of Mayhill Homes, is one of the most functional I've seen.

Instead of simply installing a wood stove in his new house, Odegaard in essence built his house *around* the stove. And this is no exotic research project— it's available now as one of Mayhill's factory-assembled homes.

The Støvehaus (see photos on following pages) uses basic heat-distribution theory, some of it researched at our own Chestnut Mountain research home, which is also a modified Mayhill home. Warm air from a centrally placed stove rises to a lofty, peaked ceiling. This acts as a plenum, collecting and channeling the air to an intake near the peak. A blower then distributes it via a housewide duct system.

The simple system works. When I entered the Støvehaus in February, a fire had been burning in its small stove for only an hour, and the blower hadn't yet kicked on. But the house was comfortable. Checks in various rooms did show temperature variations, so we started the blower manually. An hour later, there were no large temperature differentials, and rooms far from the stove were warming up nicely.

Now that the Støvehaus has proven itself in winter weather, Mayhill has put two models on its regular production schedule. For more information, write Mayhill Homes, Box 1778, Gainesville, Georgia 30503.

143

144

**Veranda-encircled Støvehaus** (top) has a cozy, rustic look. But modern second-story balconies and unique integral chimney cap at the tip of the peaked roof adapt the house to contemporary setting. This pyramid-shaped roof is more than decorative. It funnels warm air that rises through a dramatic open stairwell (lower left, facing page) into a ceiling air intake (visible above, left of the insulated stovepipe). The clever air-flow design allows waste heat from lights and appliances to be chan-neled upward along with heat from the small Fisher stove. A larger stove was originally installed, but provided too much heat for the 2400-square-foot house. The stairway opens (lower right, facing page) into the loft room that bridges the entry (see diagram, page 146). Thermostat at left of door controls blower that distributes air through ducts. If temperature drops below comfort level, a standby gas furnace kicks on. The factory-built Støvehaus is "panelized": It comes with preassembled wall sec-tions in panels, and precut floor and roof. (Photos copyright 1980 by E. Allen McGee, Atlanta.)

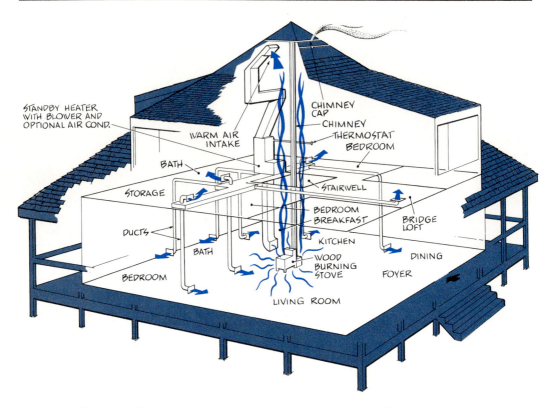

**Wide-open floor plan** on first level lets stove warm nearby seating area directly, as well as heat house via ducts. Designer put stove in central entry hall for efficiency—and to discourage furniture clusters close to the stove. (Drawing by GENE THOMPSON)

# Wood-stove cottage

*The air-flow pattern is effective for both heating and cooling*

## By Al Lees

This ruggedly styled, two-story retreat was especially designed to be heated by an efficient wood stove. And its compact floor plan (see diagram) contains still other energy-saving features. Created for *Popular Science* by architect Samuel Paul—who calls it The Timbers—the house is shaped to funnel the stove's heat upward to the second-floor bedrooms. As shown in the interior view, wood louvers across the top of this wall let the heated air flow into these rooms and the upstairs bath.

In the summer, this flow pattern can be as effective for cooling. Just open the projecting casements over the window seats, plus the upstairs windows on the east wall, and you'll send breezes from west and south up through the louvers. In most vacation areas, this feature should make air conditioning unnecessary.

And in many climate zones, your wood stove could heat the entire house. For example: With design temperatures of – 15 degrees F outside and 68 degrees F inside, a No. 4 or No. 6 Combi-Fire stove by Jøtul could supply the 45,000 Btu required. An even more efficient heating system might combine such a stove with electric-resistance baseboard that would provide supplemental heat at extreme temperatures and when the house is vacant.

If you're not into wood as a major heat source, replace the stove with a stove-fireplace; this will contribute Btu to supplement a central furnace—space for which is provided in the mudroom.

Use of glass in this house has been considered for its effect on heat loss. The northern exposure has limited glazing, while the walls to the east and south are more open, with sliding glass doors giving onto the decks.

147

**FIRST FLOOR PLAN**

Labels within first floor plan:
- 37'-2"
- DECK
- bench
- cl
- FOYER
- cl
- BED RM 11'-6" x 12'-2"
- sl. gl. dr.
- DECK
- bench
- LIVING RM sloped ceil. 17'-4" x 21'-4"
- HALL
- BATH
- 35'-0"
- window seat
- up
- htr
- w.h.
- access to crawl space
- wood stove
- stor.
- window seat
- sl. gl. dr.
- ref
- s
- dw
- w
- d
- mud rm
- KITCHEN 11'-6" x 13'-0"
- shr
- DINING
- DECK
- bench

**SECOND FLOOR PLAN**

Labels within second floor plan:
- roof
- cl
- cl
- BED RM 11'-6" x 12'-2"
- sl. gl. dr.
- balc
- upper part of living rm
- roof
- BATH
- HALL
- htr. flue
- flue
- cl
- cl
- lin
- dn
- roof over mud rm.
- BED RM 11'-6" x 11'-0"

That welcoming deck across the front leads to an entry that is recessed from winter drafts. The big living room with its sloping ceiling is the social center of the house. Two window seats supplement the central furniture grouping.

The adjacent kitchen has generous eat-in space, and beyond it is the mudroom with an auxiliary entrance, a shower, and laundry equipment.

There's a full bath downstairs, and a bedroom with its own sliding glass door to a private deck. Directly above this area is an identically sized room with its own balcony. You take your choice as to which becomes the master bedroom—weighing accessibility against privacy. Also, in the winter you may want to close off that back corner and move up to the snug rooms tucked under the ridge.

For information on obtaining construction blueprints for the wood-stove cottage, write to Homes For Living, Inc., 107-40 Queens Boulevard, Forest Hills, NY 11375. (Specify "Timbers Plan.")

# HOW TO AVOID HEAT LOSS

6

The best heating system in the world is of little value if your home can't contain the heat it generates. Heat leaks out in two ways—through the walls, ceilings, and floors by conduction, and through open cracks and crevices by infiltration.

By now, most homeowners have upgraded their insulation to some extent. But in most homes we visit, there are numerous areas that are overlooked, even if the job was done by a pro.

Hardly anyone needs to be sold on the value of adequate home insulation today, but it's important to understand something about the various types of insulation available and about how it is applied. If you find your home needs more attention in this area, you can likely do much of the work yourself— you'll see how in this chapter. Even if a pro does the job, follow behind to be sure that no details such as blocked vents are left. You may also find areas that were overlooked, such as basement foundation walls. No matter how much you've already done, chances are good that you will find something here to help you keep from heating the great outdoors.

Thanks are due *Homeowners How-To* for making available the material found in "Insulate to reduce your fuel bills" and "Listening for air leaks."

# Upgrading your insulation

*Controversies still rage as to which type, how much, and who installs it, but new studies are about to confirm basic standards*

## By Al Lees

On 500 acres of rolling countryside in Granville, Ohio the major research facility in the insulation industry is situated. It bristles with new construction; shelters as vast as airplane hangers are still rising on the hillsides. There are 728,000 square feet of labs, libraries, data banks, and supporting structures. There are more than 900 employees here—half of them scientists and engineers.

I toured the facility to learn what the home-insulation industry is doing to come up with facts to support current beefed-up standards. Too much confusion now exists, as homeowners are quoted different figures by a dozen different sources in industry and government—and conflicting advice about which insulation is the best where, and whether a do-it-yourselfer should tackle the installation.

I can't believe that any reader of this book north of the Gulf States and Southwest hasn't already upgraded his home's insulation and weatherstripping, but those few who didn't are doubtless burning their back issues in their wood stoves, right now, to make it through without a second mortgage to cover heating costs. Most of us retain a nagging suspicion that we could and should do more to lower our fuel bills.

Two developments put insulation back in the news in recent years. The long-awaited National Energy Act finally passed (in diluted form), offering homeowners a three-pronged incentive toward thermal efficiency:

- A program requiring local utility companies to offer *energy audits* of their customers' homes, and to assist them in arranging financing for whatever upgrade the audit shows is needed (mostly this will be more insulation, better weatherstripping, double or triple glazing).

- A $5 billion program of federally supported home-improvement loans for energy conservation (up to $2,500 for qualified families).

- A tax credit of up to $300 on such installations, through 1985, on any home built before April 1977 (leisure or second homes excluded).

152

The other significant development on the insulation front was the long-awaited publication of the revised and enlarged second edition of the booklet that became the bible of home insulation: the National Association of Home Builders Research Foundation's *Insulation Manual.* Since the original manual established accepted practices for installing insulation for the nation's home builders (and was much cribbed from by shelter magazines), it was widely influential.

The original manual, published in 1971, was ahead of its time in urging energy efficiency in home building; this was well before the oil embargo of October '73, from which most people date our energy crisis (wrongly—the embargo only dramatized the finite nature of the fossil fuels we'd become too dependent on, and shamed us with our profligate use of them). The second edition helps you figure the right amount of insulation for a given house in a given location (with weather data for hundreds of U.S. cities). It reprints ASHRAE's* list of R-values for building materials, insulations, and air spaces, and provides detailed sketches of proper installation methods for most types of insulation. It endorses no standards [the National Association of Home Builders'—NAHB—guidelines included in our table on page 154 are from another publication]. It simply presents the facts.

There are dozens of statistical tables of heat-gain or -loss values for windows, doors, ducts, etc., and there are work sheets for calculating a home's heating and cooling loads. Data to help you figure actual cash savings you can expect from an energy-conserving home is included. You'll find a persuasive comparison of an average house (R-13 in ceiling, R-11 in walls framed 16 in. OC, single glazing) with an energy-saver (R-30 in ceiling, R-5

foam sheathing added to R-11 batts in walls framed 24 in. OC, R-19 at floor's band joists, double glazing) that cuts heat loss *in half.*

The manual also details other ECT's (Energy-Conserving Techniques, such as weatherstripping and proper siting and sizing of windows). In short, it's an expert summary of what we know about building energy-efficient homes.

But what took me to Granville is what we *don't* yet know, or at least haven't authoritatively established with documented research: Are all these techniques for constructing new homes and retrofitting old ones truly effective, or an overkill pressed on us by the insulation industry, as some (*Consumer Reports* among them) have charged?

There is now a fairly basic agreement between major sources as to recommended R-values (see table, next page). In adjusting to a degree-day gauge, rather than the necessarily cruder and less-precise zoned map it published so widely, Owens-Corning Fiberglas (OCF) has revised its initial figures somewhat downward—or at least made them more flexible. Cost-effectiveness is the key. Farmers Home Administration (FmHA) uses a 33-year payback period (the life of its mortgages) against the NAHB's seven-year period—the average time a family keeps a home before moving.

These figures—the recommended heat-flow resistance to be built into a home's ceilings, walls, and floors—should be taken as guidelines only. To use the table, you need to know the annual heating degree-days for your area. You can obtain this from a local weather station; the NAHB manual; or from a Zip Code Data Base booklet, compiled at Granville. (It also lists latitude and longitude, average annual sunshine, and maximum and minimum temperatures for each zip-code area. For a copy, write Owens/Corning Fiberglass Laboratories, Granville, Ohio 43023.)

---

*American Society of Heating. Refrigeration, and Air Conditioning Engineers

# R-values recommended by four major sources

| Degree Days | OCF Zones | Ceilings | | | | Walls | | | | Floors | | | |
|---|---|---|---|---|---|---|---|---|---|---|---|---|---|
| | | OCF | FmHA | Prop. FHA | NAHB | OCF | FmHA | Prop. FHA | NAHB | OCF | FmHA | Prop. FHA | NAHB |
| Above 7000 | 1,2 | 38,33 | 38 | 38 | 21-35 | 19 | 19 | 19 | 12-18 | 22 | 19 | 19 | 16-22 |
| 6001-7000 | 2 | 33 | 38 | 30 | 21-34 | 19 | 19 | 19/11 | 12-17 | 22 | 19 | 19/11 | 16-20 |
| 4501-6000 | 3 | 30 | 30 | 30 | 21-30 | 19 | 19 | 19/11 | 12-16 | 19 | 19 | 19/11 | 13-19 |
| 2501-4500 | 3,4 | 30,26 | 30 | 30/22 | 19-30 | 19 | 19 | 19/11 | 12-16 | 19,13 | 19 | 19/11 | 7-19 |
| 1001-2500 | 4,5 | 26 | 22 | 22/19 | 19-22 | 19,13 | 11 | 11 | 12 | 13,11 | 11 | 11/10 | 0-15 |
| 1000 & under | 5 | 26 | 19 | 19 | 19-21 | 13 | 11 | 11 | 12 | 11 | 11 | 11/0 | 0-9 |

**NAHB guidelines** require a maximum payback of seven years (current payback is often four or five) so are lower, to trim builders' initial investment.

## WHAT SHOULD YOU TACKLE?

Once you've decided exactly how much insulation to put where, you can quickly determine whether the installation is a do-it-yourself job or will require a contractor. This decision depends on where and with what you've chosen to insulate: If you're beefing up insulation in an exposed attic floor, you can do the job as effectively as a pro, whatever your choice of material: loose fill, batts, or rolls. The underside of floors over crawl space or unheated basement or garage is similarly easy for a novice to insulate. Side walls, however, are another matter. You can glue rigid foam board on the interior surfaces (especially on foundation masonry, when finishing a basement); you'll have to cover it with sheetrock to meet fire codes, since both styrene and urethane are flammable.

But framed walls of a finished house aren't a do-it-yourself retrofit. Whether blowing in wool or adding foam-board sheathing prior to a re-siding job—take care to select a qualified contractor (see box).

As the five photos on page 156 indicate, a pressure can of foam sealant can make an important contribution to do-it-yourself insulating against heat loss. A public utility in the southwest recently studied air infiltration in 30 typical homes and traced it to these spots: baseplates, 25 percent; wall outlets, 20; ducts, 13; windows, 12½; vent hoods, 6; fireplaces, 5½; entry doors, 4½; dryer vents, 3; patio doors, 2; all other, 3½. Sealing all electrical outlets and window and door frames saved $177 in annual fuel costs. In contrast, beefing up ceiling insulation from R-19 to R-38 (six inches to 12 inches) saved only $78.75, and increasing wall insulation from R-11 to R-19 saved only $55.

# One homeowner's way of choosing how and who to do it

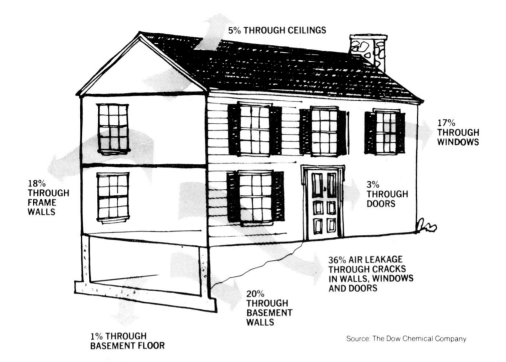

5% THROUGH CEILINGS

17% THROUGH WINDOWS

18% THROUGH FRAME WALLS

3% THROUGH DOORS

36% AIR LEAKAGE THROUGH CRACKS IN WALLS, WINDOWS AND DOORS

20% THROUGH BASEMENT WALLS

1% THROUGH BASEMENT FLOOR

Source: The Dow Chemical Company

Comparative heat-loss figures, above, are for a typical two-story home with moderate insulation, double-glazing, and 2,000 square feet of opaque wall area. By adding inch-thick Styrofoam TG panels to sidewalls and foundation—roofline to frostline—says Dow, you can save 24 percent on winter fuel costs: You trim the three highest loss areas.

These claims impressed David Curtis of Rochester, New York. He planned to have his home re-sided, and had studied all other retrofit options.

"Since my walls already had 2-inch batts, I feared that any blown-in material might get hung up on the existing insulation." He finally chose the sheathing envelope of Styrofoam because it can maintain its R-value over a long period. He also wanted insulation that transmits some moisture vapor.

"I wanted to avoid condensation in the wall, which can result in wood rot. Styro-foam has a moisture transmission factor of 0.6—ideal under new siding. I also liked its t&g edges for a continuous envelope to cut air infiltration." He chose Tedlar siding because "it will look the same after 10 years of weathering. In upgraded appearance, comfort, and fuel economy, the whole job is already worth what I paid. And its value will gain along with that of the house itself."

But Curtis reached this happy state through careful choice of materials and professional help. After reviewing a list of 30 home-improvement contractors in his area, he asked the top eight for job estimates (anyone who subcontracted work was eliminated). Each proposed to do the job differently, so Curtis had to educate himself to determine which was good for him. He asked for references from past jobs, and checked them out to find which best matched his own requirements.

155

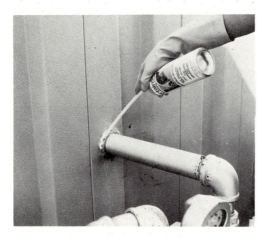

**Air infiltration** can account for a major portion of a home's heat loss, as the diagram on preceding page indicates. Exterior leaks—around doors, windows, and pipes, or between dissimilar materials—can be stuffed with scraps of mineral wool and weather-sealed with caulking. But now there's a way to get an even more effective insulating seal: Use polyurethane foam in a can. It cures into a semi-rigid, water-repellent bead that dries tack-free in 20 minutes. One inch rates R-5. It's available in most building supply and hardware stores from several manufacturers.

**Re-siding is ideal time** to beef up thermal efficiency of walls with layer of Styrofoam between old and new. Long nails pass into original substrate.

So leak-sealing can be effective, whether you do it with an expandable foam or conventional caulking. The latter has no real insulating value, however; it just plugs the leak. Movable panels, such as windows and doors, can't be sealed, of course. For those, you need to make a choice of weatherstripping of doors or windows.

Don't confuse these "canned" insulations or rigid foam boards with urea-formaldehyde (UF) foam, which has been used in the past for foaming walls in existing homes. UF foam, now banned nationally, is an excellent insulation—but the curing process is critical, and if an improper cure results it can cause the release of formaldehyde gas into the living area. Research into the effects of exposure to this material is still going on.

Are we certain all this is effective? All the statistics I've quoted thus far were based on actual monitoring of homes. But

I suspect the last word—and *the* authoritative statistics—will issue from the OCF Technical Center in Granville. Photos on the next page show two of the facilities: First, there are the huge hinged sections of a calibrated hot box to let you mount an entire wall section in a 9-by-14-foot frame that divides the box (when it's pivoted closed and locked) into exterior and interior sections. As the temperature on the "exterior" of the wall is varied to simulate changing weather, many thermocouples (heat sensors) measure the effects on the "interior" temperature. By altering the framing, insulation, and sheathing of the test sample, various combinations can be compared. Windows and doors can be built into samples to test their thermal efficiency. Compressed air can be fed into the cold box to simulate air infiltration.

The second facility is brand new; calibration has just been completed. It does much the same job in measuring *vertical* heat flow, and on an even larger scale. Complete floor or roof assemblies can be built over the below-grade pit, which represents the interior of the house. Weather conditions can be simulated within the giant test chamber (it measures 25-by-35-by-75 feet and cost over $1 million).

In 1979 The Department of Commerce proposed criteria to accredit labs such as this one under the National Voluntary Laboratory Accreditation Program (NVLAP), "to encourage manufacturers of thermal insulation to use the highest caliber of testing services." Manufacturers who want such accreditation for their labs have to demonstrate proper use of the thermal test methods of ASTM (American Society for Testing and Materials). In other words, Commerce is admitting that, under federal supervision, insulation manufacturers are better qualified to set standards and check quality control than the government is.

**Large-scale test capability** of OCF's hot box permits mounting of entire wall assemblies. Thermocouples measure horizontal heat flow from warm side.

We don't show a photo of the most impressive test facility. On a remote, windswept corner of this vast property, OCF technicians have erected a trio of identical three-bedroom ranch houses—each partly slab-on-grade, partly over crawl space, partly over excavated basement—built from the supertypical Herman York plan that the NAHB *Insulation Manual* still uses as the "representative dwelling" in its heat loss (and summer heat gain) examples.

The trio makes an eerie sight out there, as if an ambitious (but not very creative) developer went bankrupt, leaving these cookie-cutter homes in bleak isolation from any sense of community. The homes are sealed for a full year's monitoring and won't be opened until winter's end.

Identical? In appearance only. The first house has *no* insulation—simply empty cavity walls and a bare attic floor. The second has six inches of fiberglass in the ceiling and three inches in the walls—the standard of well-built homes prior to the energy crisis, and still the HUD minimums for Ohio. The third is fully insulated to current beefed-up standards: R-19/38/19. In order to get those R-19 walls, one variation in construction had to be introduced: They're framed with 2x6's.

**Even greater capacity** of vertical test chamber allows heat-flow measurements through entire floors (photo) and roofs (inset sketch) suspended over pit.

The homes are identically finished and furnished inside; every room bristles with thermocouples and watt meters to measure temperature distribution and the power requirements to maintain it at a preset standard. A nearby weather station charts the changing atmospheric conditions to provide researchers with detailed correlation between weather and the thermal performance of each house. Electric resistance heaters are used, but cooling, in a recent summer, was done with chilled brine, for more exact measurement than would be possible with air conditioners or heat pumps. There are four- to six-foot ground probes around each foundation to measure subgrade heat loss; 900 instrumentation points inside monitor air infiltration, attic heat buildup, heat loss and gain through windows.

The data base—all filed on magnetic tape—is so complex that it will take the center's vast computer system months to translate it into meaningful terms, after the program is completed. But once that's done, those terms will be definitive. It would be presumptuous indeed for any of us, including government regulatory agencies, to argue with such field-research readout.

But surely somebody's peeked along the way? I asked Len Stenger, the project manager, if there had been any surprises thus far. He couldn't suppress a satisfied smile, so I suspect that OCF's current insulation standards are proving to be right on target.

# Year-round energy miser

*New windows, passive heating and cooling make this new house a money-saver*

## By V. Elaine Smay

What would happen to the temperature in your house if you left it unheated from October to March?

Unless you live in the Sun Belt, chances are it would get pretty chilly inside. But this house in Frederick, Maryland, was unoccupied and unheated all one winter. "For five weeks the outside temperature dropped into the teens at night and rose into the 20's during the day," says Michael Milliner, president of the company that built the house. "The lowest it got inside was 56 degrees, the average was 60, and the most it varied in 24 hours was two degrees."

Obviously, this is no ordinary house. It is snugged into a south-facing hillside and blanketed with 18 inches of earth, which shelters it from the temperature extremes of the air. Most of the windows face south and serve as passive solar collectors. Company calculations indicate that the sun will supply 75 percent of the already low heating needs of the 2,300-square foot house. The remainder, only 1.3 Btu/degree-day/square foot, compares with seven to 10 Btu/degree-day/square foot for the average modern house. "Burn less than a cord of hardwood in the air-tight wood stove, and you've got it," says Milliner.

The alternative backup heat source is a heat pump, which can also supply air conditioning—if needed. It may not be. A third of the cooling load will be supplied by a heat-pump water heater. Passive cooling is also part of the design. An 80-foot-long, 12-inch-diameter concrete pipe is buried at an average depth of 10 feet on the north side of the house. Air drawn in through the pipe will average about 15 degrees below ambient temperature, calculations indicate.

## HIGH-R WINDOWS

The greenhouse and clerestory windows consist of a four-layer sandwich that "could revolutionize passive solar heating," says Milliner with enthusiasm. The two outside layers are glass, but the inner layers are a new film developed by 3M. The sandwich adds up to an R-value of 3.9 (compared with R-1.88 for standard double-glazed windows). "These windows outperform double-glazed windows that are exposed eight hours a day and covered with R-6 insulation for 16 hours," Milliner says.

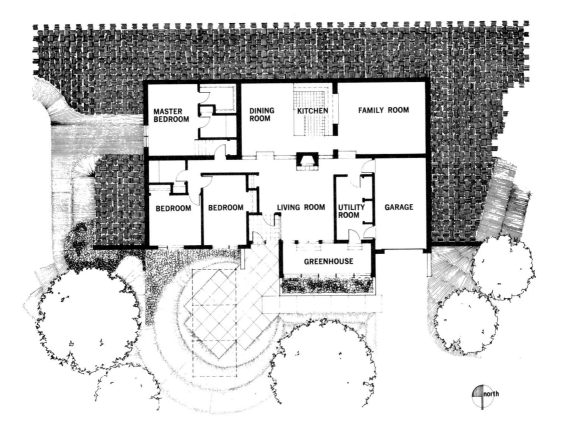

MASTER BEDROOM · DINING ROOM · KITCHEN · FAMILY ROOM · BEDROOM · BEDROOM · LIVING ROOM · UTILITY ROOM · GARAGE · GREENHOUSE

north

The walls of the house are of 12-inch reinforced concrete block, and the roof is made of hollow-core precast-concrete planks. The house is waterproofed with Bentonite, an expansive clay, and is heavily insulated on the outside, which lets the concrete serve as thermal mass to store heat and even out diurnal temperature swings.

The house won both a design award and a construction grant in the HUD/DOE Cycle 5 Residential Solar Demonstration Program. Robert May was the architectural consultant, Michael Tallmon was the solar designer, John Darnell served as a consultant, and Robert Whitesell was the structural engineer. M. S. Milliner Construction, Inc., sold the house and its 3½-acre wooded lot for $144,800. "People are going to have to bite the bullet up front for a passive-solar earth-sheltered house that is properly built," Milliner says. "But they will not only save energy. This house requires virtually no exterior maintenance, isn't going to burn, termites and rot won't touch it, and it's virtually stormproof."

The company (302-A E. Patrick St., Frederick MD 21701) sells construction drawings for about $100 a set, and 30 slides of the design and construction, along with a synopsis of the procedures, for about $30.

**On a winter day,** the sun shines through the clerestory windows, providing direct solar gain. The greenhouse can provide either direct or isolated solar gain: For direct gain, the sliding glass doors between the living room and greenhouse are left open; for indirect gain, the doors are closed. The heated air in the greenhouse rises, passes through vents to the hollow cores of the concrete roof panels, and flows into the rooms at the back of the house. The warm air gives up some of its heat to the concrete mass. Cooler air near the floor is channeled back to the greenhouse through ductwork (not shown) to complete the convective loop. If necessary, the air handler of the heat pump can be used to boost air flow. At night, the heat stored in the concrete is released to the rooms. The wood stove or heat pump can provide backup heat if needed. Earth-tempered (about 50 degree) makeup air, which may be needed when the wood stove is used, can be drawn in through a buried pipe (see text).

In summer, the earth pipe cools and dehumidifies incoming air, which enters the house through the ducts, rises as it warms, and exits at high vents behind the chimney. If necessary, an exhaust fan in the chimney or the air handler can be used to boost air flow. Or the heat pump can provide air conditioning. A thermostatically controlled fan exhausts hot

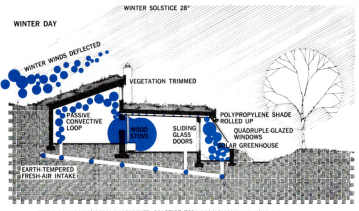

WINTER DAY

WINTER SOLSTICE 28°

WINTER WINDS DEFLECTED

VEGETATION TRIMMED

PASSIVE CONVECTIVE LOOP

WOOD STOVE

SLIDING GLASS DOORS

POLYPROPYLENE SHADE ROLLED UP

QUADRUPLE-GLAZED WINDOWS

SOLAR GREENHOUSE

EARTH-TEMPERED FRESH-AIR INTAKE

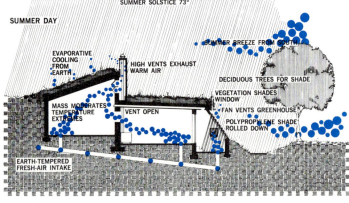

SUMMER DAY

SUMMER SOLSTICE 73°

SUMMER BREEZE FROM SOUTH

EVAPORATIVE COOLING FROM EARTH

HIGH VENTS EXHAUST WARM AIR

DECIDUOUS TREES FOR SHADE

VEGETATION SHADES WINDOW

MASS MODERATES TEMPERATURE EXTREMES

VENT OPEN

FAN VENTS GREENHOUSE

POLYPROPYLENE SHADE ROLLED DOWN

EARTH-TEMPERED FRESH-AIR INTAKE

air from the greenhouse; a polypropylene-mesh shade blocks 80 percent of the sun. The cool earth around the house acts as a heat sink.

# Super-insulated houses

*Airtight double walls mean near-zero heat loads*

## By Daniel Ruby

The conventional-looking house pictured here has very unconventional fuel bills. During an entire icy Saskatchewan winter, Peter and Judy Fretz's house used only as much fuel as a standard house would have needed in two weeks.

The secret is under the skin: 12 inches of insulation stuffed into double exterior walls; greatly increased insulation in the floors, ceilings, and foundation; a tightly sealed vapor barrier; and vestibule entries.

The Fretz house is one of a relative handful of homes in the U.S. and Canada that are pointing the way to a new trend in building. Super-insulated houses—also called low-energy or conservation houses—approach home heating by cutting energy demand instead of increasing supply. People who have built and lived in them believe that they perform better and cost less to build than

**The Fretz house walls,** two sets of 2x4's separated by plywood plates and sheathing, were constructed in sections, then raised into place. Continuous vapor barrier covers inside of plywood. Windows are concentrated on south; overhangs prevent overheating in summer. Fretzes modified the Cape Cod design with help from Saskatchewan Research Council.

houses with complicated active or passive solar systems.

In fact, super-insulation is a kind of passive solar design. Multiple-glazed windows are concentrated on the south side, and direct solar gain is an important source of heat. But, because heat travels both ways through windows, much less glass is used than in most passive homes. Also, no special attention is given to adding heat-storage mass.

Instead, dramatically increased insulation and decreased air leakage seals in the modest solar gain and the internal heat given off by human bodies and electrical appliances. The result is an energy saving of at least 75 percent over a house with average insulation (see chart). A

similar saving results in summer: By keeping hot air out, air-conditioning cost approaches zero.

Some super-insulated houses can even get by without a backup heating plant. "That's the quantum jump," says Harvard researcher William Shurcliff. "That $4,000 saving makes up the cost of the other features."

So far, few builders have been confident enough to eliminate the furnace, and most super-insulated houses have cost three to eight percent more to build than homes with average insulation.

The slight capital-cost penalty does not dampen enthusiasm, though. "Within a few years, super-insulation will be the commonplace of building," says Wayne Shick, a retired University of Illinois architect whose 1974 Lo-Cal House design is credited with starting the super-insulation boom.

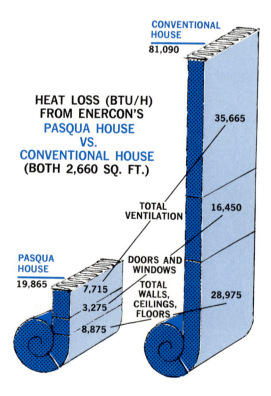

HEAT LOSS (BTU/H)
FROM ENERCON'S
PASQUA HOUSE
VS.
CONVENTIONAL HOUSE
(BOTH 2,660 SQ. FT.)

CONVENTIONAL HOUSE
81,090

35,665

TOTAL VENTILATION    16,450

PASQUA HOUSE
19,865        7,715    DOORS AND WINDOWS

3,275    TOTAL WALLS, CEILINGS, FLOORS    28,975

8,875

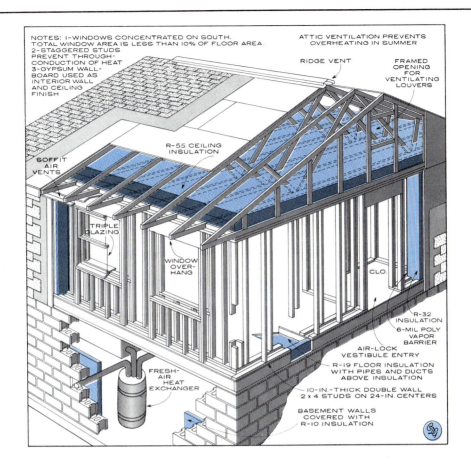

NOTES: 1—WINDOWS CONCENTRATED ON SOUTH. TOTAL WINDOW AREA IS LESS THAN 10% OF FLOOR AREA. 2—STAGGERED STUDS PREVENT THROUGH-CONDUCTION OF HEAT 3—GYPSUM WALL-BOARD USED AS INTERIOR WALL AND CEILING FINISH

ATTIC VENTILATION PREVENTS OVERHEATING IN SUMMER

RIDGE VENT

FRAMED OPENING FOR VENTILATING LOUVERS

R-55 CEILING INSULATION

SOFFIT AIR VENTS

TRIPLE GLAZING

WINDOW OVER-HANG

CLO.

R-32 INSULATION

6-MIL POLY VAPOR BARRIER

AIR-LOCK VESTIBULE ENTRY

R-19 FLOOR INSULATION WITH PIPES AND DUCTS ABOVE INSULATION

FRESH-AIR HEAT EXCHANGER

10-IN.-THICK DOUBLE WALL 2 x 4 STUDS ON 24-IN. CENTERS

BASEMENT WALLS COVERED WITH R-10 INSULATION

## Pieces of the puzzle

Typical insulation values and other features of a super-insulated house are shown in the drawing, but designs vary widely. "The key to a cost-effective design is striking a balance among the many elements," says David Robinson, a Honeywell development engineer who devised mathematical techniques to help designers. "If you are going to use only double glazing in the windows, for example, then it's a waste of money to put more than R-21 of insulation in the walls." The required R-values will determine what kind of wall construction is needed. Robinson's guidelines also indicate that you should spend the same amount of money on energy-saving in-stallations as you'll spend on fuel over the life of the house. But his methods don't yield easy decisions on some other features. Air-lock doors? The experts are split. They also dispute the importance of a perfectly sealed vapor barrier. Wayne Shick says a barrier is important mainly to protect insulation from humidity; perfect seal is less cost-effective than using superior windows. Leland Lange says sealing is crucial. "Upward of 40 percent of the energy load in an average house comes from heating infiltrated air." If sealing drops air-change rate to under 0.2 an hour, a fresh air heat exchanger may be needed to combat indoor pollution.

After Shick published his design in 1976, other academic and scientific groups took up the crusade. The Saskatchewan Research Council built and monitored a demonstration house that incorporated many of Shick's ideas. The same group is now sponsoring a "parade" of energy-efficient homes in Saskatoon.

But the breakthrough came when a number of independent owner-builders, including the Fretzes and Eugene Leger applied the concept in actual homes.

"I call those people the granola boys," says Harry Hart, a commercial builder in Lorton, Virginia. "Now it has gone beyond that stage." Hart has developed two subdivisions of houses based on the Lo-Cal design, and Enercon Builders, in Regina, Saskatoon has sold 70 super-insulated homes and licensed its design to other builders.

"Business is flying," says Enercon's Leland Lange. "In our area, all sectors of the industry—from the subtrades to the financial institutions—are getting involved."

Harry Hart is not so sanguine. He says that only about 20 percent of his buyers were sold by the energy efficiency of his houses. "The consumer is a funny cat. Many buyers say they are interested in energy, but they base their decisions on items like fancy bathroom fixtures."

"Energy awareness in building may have to work its way from the north down," comments Shick. "But as energy prices soar, there will be a moment of awakening."

While super-insulation backers are universally excited about the concept, specific designs vary. Some favor double walls with 8½ inches of insulation, some 12 inches. Others get by with single 2x6 or 2x8 walls with exterior sheet insulation supplementing standard fiberglass batts. Some require an airtight seal, heat exchangers and vestibule entries. Others say these aren't needed. There are many pieces to the super-insulation puzzle. How they go together makes a big difference (see box, page 165).

Right now, many super-insulated houses are rather plain looking. But this is a result of builders and buyers trying to reduce cost, not of design limitations. Within certain constraints, such as the number of windows and overall size of the house, super-insulated houses could be architectural eyecatchers.

Will they replace other types of solar design? Shurcliff says there will still be a role in very sunny areas for passive houses with huge amounts of glass and thermal storage. "But almost every kind of house has some sort of catch," he says. "With superinsulated houses, nobody has found an Achilles heel."

# Insulate to reduce your fuel bill

*Money you save by stopping conduction and infiltration will keep your heating costs under control*

## By Evan Powell

For most homeowners, it's no longer a question of whether to add insulation to their homes, but only when and how. Yet, sometimes, adding insulation is wasteful. And if the insulating is improperly done, it can be less effective than what was there before—it could even be hazardous.

If you check with five materials suppliers, you'll get five different answers as to the best way to do the job. Actually, they could all be right—but they could all be wrong, too. Sorting out available information and combining it with my experience was the major thrust of the research for this story. These facts soon became evident:

- In most cases, insulation offers the best potential for reducing home operating costs.

- In many of the most important areas, you can save money and, often, get a better job, by doing much of the work yourself.

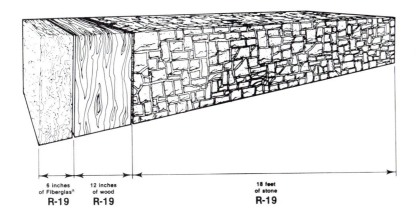

6 inches
of Fiberglas®
R-19

12 inches
of wood
R-19

18 feet
of stone
R-19

**How much difference** does it make to insulate your home? As you can see in the diagram (left), six inches of Fiberglas insulation with an R factor of 19 is equivalent to a 12-inch thick wood wall, or a stone or brick wall 18 feet thick.

167

**Adding insulation** is relatively easy in accessible attics. Loose fill or fiberglass batts are probably easiest for the do-it-yourselfer to handle. Added insulation must be unfaced so that moisture will not be trapped between vapor barriers. Note protective gear.

- Every common insulating material has advantages and disadvantages. Your home may require a combination of two or more to get the best job.
- Proper installation is often as important as the type of insulation used.

## HOW INSULATION WORKS

Between the uninsulated exterior and interior walls, or between roof and ceiling of a typical home is a combination of timbers, sheathing and air space. Units of heat always travel from a warmer to a colder surface. This means that, in winter, heat tries to escape from inside the home to the outside, while in summer, the opposite is true.

One escape route is conduction. For example: Heat units strike the exterior wall surface, usually a material that conducts heat easily, and though they may slow up briefly when they enter the air space between walls, they soon are conducted along by eddy currents. Once they reach the interior wall surface, they penetrate into the room rather easily.

Heat also enters or leaves the living area through infiltration. The typical home has thousands of cracks, crevices and spaces through which air can pass.

Insulation reduces both these actions by filling wall cavities with millions of "dead ends." The material blocks stray air currents that manage to enter the space, and heat units have to seek a way through a maze of air spaces to work their way out. Heat loss is slowed so much that the effect of added insulation is immediately apparent in reduced furnace or air conditioner running time and more constant temperatures.

Insulation also affects latent heat—the kind you can't measure with a thermometer. If you stand next to an uninsulated wall during a cold period, you may feel a "wash" of cold air off the wall, even if the room temperature is 70 degrees; in summer, in the same spot, and

with room air at the same temperature, you might feel warm. Such heat or cold plays a big part in determining comfort, which in turn leads to adjustment of the air conditioner or thermostat—and increased energy usage. Insulation and ventilation help prevent these effects.

An insulating material's ability to block heat flow is commonly expressed by an "R" factor, a simple measurement of thermal resistance. You can use this factor to compare the effectiveness of two insulating materials of equal thickness, or you can compare materials by determining the R factor per inch of each (see chart).

### Insulating materials

| Material | Typical "R" value per inch |
|---|---|
| Rock wool batts and blankets | 3.1 to 3.6 |
| Rock wool loose fill | 2.7 to 3.2 |
| Glass fiber batts and blankets | 2.7 to 3.7 |
| Glass fiber loose fill | 2.1 to 2.4 |
| Cellulose loose fill | 3.1 to 3.7 |
| Polystyrene (expanded) | 4.0 to 5.26 |
| Polyurethane | 6.25 |
| Vermiculite | 2.1 to 3.0 |
| Amorphous silicate | 4.0 |

## TYPES OF INSULATION

The most popular home insulating materials are glass fiber and mineral wool. Glass fiber is made from molten silica or sand that is "spun," much like cotton candy, into strands. Mineral, or rock, wool is similar, but it's spun from slag (metal-refinery waste). Both are available in (a) batts—matted strands several feet long, usually in widths to fit standard, on-center, wall-stud spacing (16 or 24 inches), or (b) blankets—16- or 24-inch wide rolls from which appropriate lengths are cut.

**New, loose-fill** Dacotherm (right is an amorphous silicate material that's recommended by manufacturer for use as both new and add-on attic insulation; it has an R factor of 4, and is available in most eastern states. Such materials as Insta-Foam's Froth-Pak kit or Coplanar's Polycel allow you to caulk with insulation (below, left). Foams are especially suited for sealing around window and door jambs, sills and headers of an unfinished wall. After curing, foam should be covered for maximum fire protection. Rmax Thermawall (below, right) is an insulating drywall sandwich panel, which is made up of a foil-backed rigid foam sheet that is permanently bonded to gypsum board. It's designed for use inside perimeter walls and also for cathedral ceilings. Thermawall can be applied either to a conventional dry wall channel, or with its own fastening system as is shown in the illustration.

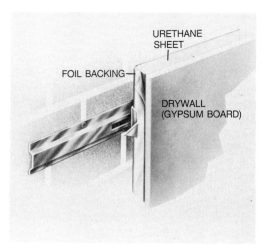

Sometimes, glass fiber and mineral wool are chopped into loose-fill insulation. So are such other materials as cellulose and vermiculite. Loose fill is usually packed in bags. It can be poured into cavities between studs in house framing, or blown into place with a machine.

Cellular foams are also important insulating materials. They have a high R factor, but they're flammable so the living-area side must be protected with a nonflammable wall material, such as gypsum board. Polystyrene and polyurethane are available in boards or 4 x 8 sheets.

Moisture in walls contributes to such problems as structural rot and mildew. It also can make insulation ineffective. To prevent moisture from penetrating walls and insulation, a vapor barrier is used.

Many insulation materials are faced with plastic, or have an asphalt-coated kraft paper or foil membrane which serves as the vapor barrier. This facing must *always* be positioned toward the living area of the home. Thus, in ceilings, the facing is down; under floors, it's up. Should the barrier be torn or cut during installation, it's important to take the time to tape it back together.

Another method of providing a vapor barrier: Installing a covering of plastic film or other material on the living-area side of unfaced insulation or loose fill. Usually, the film is put on after the insulation is put in the wall.

Insulation that has a vapor barrier should never be installed over existing insulation, since moisture can be trapped between the two barriers. If your present insulation has no vapor barrier and it's impossible to add one, you can get some protection by undercoating the outside-facing walls of a room with two coats of aluminum or oil-based paint or by using a special paint designed to form a vapor barrier.

## SAFETY FACTORS

When *properly* manufactured and installed, insulation involves little danger. Still, there are some risks to be guarded against with various kinds of insulating materials.

Flammability is the first consideration. Glass fiber, mineral wool and loose fill, such as vermiculite, present little fire hazard, although the resin binders used to hold batt and blanket fibers together and the paper backing can burn. Plastic foam boards are combustible, and most building codes require that they be covered with a fire-retardant material.

Most notorious for the fire hazard associated with it is cellulose, a surprisingly good insulator that's made primarily from scrap newspaper. The danger comes from "backyard producers" who would have you believe insulating your home is just a matter of chopping and shredding newspaper and blowing into the walls or attic. The paper must be fireproofed, which usually means soaking it in a boric acid solution (sulphates have been used, but they can cause corrosion), and then drying it out thoroughly. The big-scale manufacturers have elaborate facilities for doing this, but most backyard producers simply spray the "mulch" with a solution and let it dry in the sun—which gives sparse protection, indeed. If you use cellulose insulation, be sure it carries a label stating that it meets the amended Consumer Product Safety Commission (CPSC) standard for flame resistance and corrosiveness.

There is also speculation that insulation may be increasing the risk of electrical wiring problems. All electrical wiring has some resistance to current flow and that creates heat buildup. If the wiring is trapped between two layers of insulation, the heat buildup can become

# Check windows, ventilation before you insulate

Don't make insulation the first step of your home's energy reduction plan. If you have a number of large, loose windows, modifying them could bring about energy savings equal to any insulation you might add. Checking windows and doors should be your first step.

Every part of your home's structure offers some resistance to heat flow and has its own R value. A typical single-pane window is only about 0.9, compared to an R value of about 5 for a typical uninsulated wall with wood exterior. The window area loses about five times more heat from the living area than an uninsulated wall section of the same dimensions. Siting conditions may change the picture, however. Southfacing windows may capture solar heat during the day to help offset the loss. By the same token, windows on the north side of the house may lose even more. A double-glazed window or storm window improves that R factor to somewhere between 1½ and 2½, depending on air space between the two panes. That may not seem great, but it's about twice as good as a single window.

In addition to heat loss through conduction, windows are notorious as primary areas of air infiltration. If your windows are loose enough to rattle, or if cracks are evident, your heat loss is substantial. Don't rely on "feeling" air blowing in around the window—the air may be flowing out, rather than in. A high-pressure area is created on the side of a house where the wind strikes; a low-pressure area, on the opposite side. This pressure differential is the cause of air leaks, and it's why heat loss is significantly greater on a windy day than on a still one.

The savings possible by improving windows varies greatly. If your home is relatively new with tight-fitting, double-glazed or double-pane windows, or if you already have storm windows, there's little else to do other than check the condition of caulking and weatherstripping. But if you find lots of cracks and crevices where air can seep through around loose-fitting, single-pane windows, there's a quick payback to be gained from improvements.

Installing new weatherstripping and caulking can help such a condition. Adding storm windows is even better— many tests, including some by the National Bureau of Standards, have indicated savings of $250 or more per season. If you install permanent storm windows, rather than the type you remove each summer, there may be little need to weatherstrip existing windows, since most infiltration would be stopped by the tight, permanent seal.

Storm windows can be added in steps. If wind strikes your house primarily from one direction, start with that side of the house; otherwise, start with the north side. Other possibilities: Insulated shades; "foam sandwich" shutters that allow sun in south-facing windows during the day, close them off at night.

Finally, look at the ventilation system in your home. In the attic, air should be able to flow freely in the eave vents and out a ridge, gable or vent located at or near the roof top. Without good ventilation, moisture may form and cause structural damage; it may also penetrate into insulation and reduce its effectiveness.

You should also have good ventilation in any unheated area, such as the basement. In crawl spaces, foundation vents should be provided to allow a crossflow of air and a vapor barrier of at least 4-mil polyethylene should be installed over exposed earth to reduce heat loss, and moisture problems.

**Stanley's Hinged Weather Seal,** shown at left, is one of new weather-stripping materials available for use on casement, awning, French-style and other problem windows. Use replacement weatherstripping on existing storm windows and doors where original material has deteriorated (center); after removing all worn material, press in new weather-seal and, according to Stanley, a thermal barrier will be created. Any window cracks or openings through which air can penetrate can be sealed with a premium-quality caulking (right), after all loose material is chipped away.

high enough to deteriorate the wire's insulation and exceed the temperatures at which it is rated. To avoid sandwiching electrical wiring between insulation, The National Electrical Code requires that you build a small barrier at least three inches away from any recessed lighting fixture. When insulation is used in walls, the installer should turn off the power, remove switch and receptacle covers, then clear out insulation from the area adjacent to the box.

Installing insulation has its own set of hazards. Many people suffer skin reactions to glass fiber and mineral wool or the chemicals used as fire retardants for other materials. For this reason, it's always wise to wear gloves and long sleeves when installing insulation. Use a disposable mask, available at most paint and hardware stores, to avoid inhaling dust and small particles, and wear goggles. A hard hat is helpful, especially in attic work, to avoid scratches and cuts

from protruding nails. In summer, work in the attic only in the coolest part of the day, and provide ventilation.

## INSTALLATION

Only after you've caulked, sealed and weatherstripped windows and doors and provided ventilation, especially in attic areas, are you ready to get down to the job of installing insulation. Insulating ceilings—it requires a minimum of effort and equipment—is a job you very likely can do yourself, in stages. The drawings that follow give the basic steps, but be sure to keep these rules in mind:

- Keep insulation at least three inches away from recessed lighting fixtures. If possible, replace recessed fixtures with surface-mounted ones; this not only eliminates a safety hazard, but generates more heat in the room area.

173

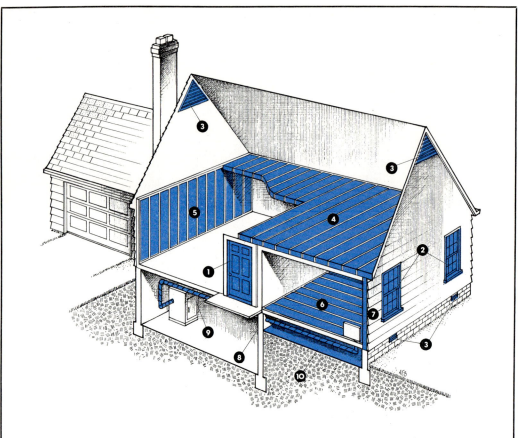

## Insulation checklist

This 10-point checklist will help you establish a long-range program to make your home weathertight. Follow the order in which items are listed.

**1.** Check caulking and sealing around doors, windows and other critical locations. Install new weatherstripping on windows and doors wherever necessary.

**2.** Install storm windows and doors, or double-glazed windows, on the north side of home and on the side that is affected by prevailing winds.

**3.** Be sure that adequate ventilation is provided in attic and foundation walls. Add vents if necessary.

**4.** Add insulation to attic area over the heated areas of home. First you will need to determine the R value of existing insulation—R factors of 25 to 30 are usually cost effective. Be sure there's a vapor barrier and take care not to block any of the attic vents.

**5.** Add insulation or insulate walls and ceilings between unheated areas, such as garage and living area.

**6.** Insulate ceilings in unfinished, unheated basement. This will keep such areas from cooling living areas.

**7.** Insulate exterior walls.

**8.** Insulate above-grade walls in finished basement.

**9.** Insulate heat ducts that pass through unfinished, unheated areas.

**10.** Install a vapor varrier over exposed earth in any crawl space.

- Don't sandwich electrical wiring between layers of insulation.
- The vapor-barrier side of the insulation must face the living area of the home. When adding insulation, use only unfaced types.
- Make a small shield to keep insulation from blocking vent openings and outlets.

You may want to keep the "cold roof" concept in mind when planning your insulation job—it's probably the best way to prevent damage from ice dams in winter. To promote a cold roof, insulate ceiling areas to an R-25 or even R-30 level to prevent heat loss from the area below. Be sure there's enough ventilation so there's an air wash between eave vents and ridge; this may involve adding open vents or, preferably, a ridge vent across the top of the roof. Snow then should melt evenly across the roof, or from the edges inward.

The R factors of insulation are additive, but there is a point of diminishing return. For most areas of the country, ceiling insulation of R-25 to R-30 and wall insulation up to R-19 is cost effective. Above that, the return may be so far into the future that's it's not worthwhile. But, if your home is insulated below the recommended levels, there are few jobs that will give you more satisfaction than improving the insulation. Not only will you enjoy the increased snugness and comfort of your home, but you'll be tickled pink by the reduced cost of operating it.

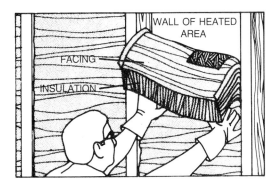

**To insulate an attic** area, first put down walkboard and be sure you have the necessary safety equipment (also make sure that you wear "safe" clothing—see text). Begin laying insulation blankets at the outer edges and work toward the center as shown at top, left. Be sure the vapor barrier faces downward—toward the living area of the house. Strip facing from the insulation and chink between the chimney and wood framing, as illustrated above. Keep insulation at least three inches away from any recessed lighting fixtures. When insulating walls between heated and unheated spaces, push the insulation into place between the wall studs, as shown at left. Allow the flange on the insulation to stick out far enough to staple it to the framing members.

WALL OF HEATED AREA

FACING

INSULATION

**It it's necessary,** make baffle to keep insulation from blocking the eave vent, as shown. Insulation should cover the top plate of the wall, but it should not extend into the eave space. These, and the drawings on the previous page, contributed by Owens-Corning, depict Fiberglas batt insulation, but the same technique would apply with insulation that is poured.

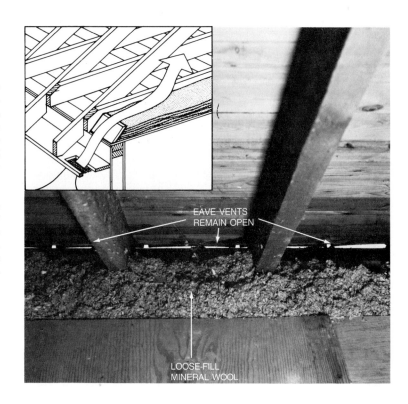

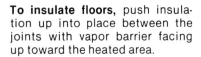

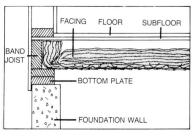

**To insulate floors,** push insulation up into place between the joints with vapor barrier facing up toward the heated area.

**Staple a network** of chicken wire under insulation to keep it in place, as shown, or use special push-in metal straps.

**Be sure the insulation** fits snugly against the band joist and that it overlaps the bottom plate.

LOOSE FILL

CHINKS

LIVING AREA

**Cavities in masonry** walls can be insulated during the construction process to double the R factor, using vermiculite-based material like Zonolite, which is treated for water repellency and requires no vapor barrier. Zonolite is also suitable for use as attic insulation.

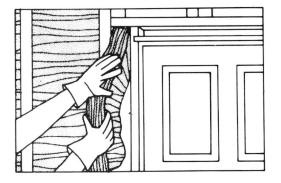

**Some final steps** in the insulation process are shown at left and above left: Use scraps to chink in cracks and small areas around doors, windows, etc. Finally, cover the insulation with wall finish.

**Blower door** mounts in any exterior door to depressurize the home and "amplify" infiltration. Here, Energy Clinic's Bill Munsell reads pressure differential between indoors and out to evaluate the effectiveness of his sealing efforts.

# All-out energy audits— How the pros spot heat leaks

---

*Leak-seeking instruments pinpoint exactly where you're wasting energy*

---

## By A. J. Hand

I picked up the instrument and peered through its eyepiece at the room. What I saw looked like a bad TV picture—in black and red. Scanning the front door, I could see a band of black flowing in over the sill. I turned to a window. Black seemed to ooze from the glass and flow down the wall.

I was looking through an infrared scanner, a sophisticated tool designed to see heat. Cold—such as the air slipping in under a front door or washing down a frigid window pane—shows up as black. Heat—in the form of long-wavelength infrared radiation—appears as red.

Such scanners are one of a number of scientific instruments that take the guesswork out of home-heat-loss calculations and point the way to cost-effective energy-saving measures. Along with blower-door depressurizers, smoke pencils, and furnace-testing equipment, they are the tools of a new but fast-growing trade— that of the home-energy auditor.

This kind of specialized contractor performs a unique service. The auditor comes into your home, makes exact measurements of its energy performance, and suggests ways you can improve it. In some cases, the auditor may actually undertake some of the recommended modifications.

A short time ago, the number of firms offering such a service—they are frequently called house doctors—could be counted on the fingers of one hand. Now they are popping up all over the country. Their services can cost the homeowner anywhere from $100 to $500 or more, depending on the sophistication of the analysis and whether or not the service includes any corrective measures. In general, house doctors will not perform complete retrofits; these can be done by the homeowner himself or by contractors at a cost of $1,000 and up.

Do you need a professional audit to pinpoint heat losses before commencing a retrofit? Not necessarily. Some kind of analysis is needed, but simply having lived in a house may tell you enough about where your leaks and drafts are. Or you could take advantage of the various free or almost-free audits that are offered. Project Conserve, developed by the Federal Energy Administration, is an example. You enter data about your home on a computerized questionnaire and mail it in for processing. In return, you get a computer printout listing possible ways to save energy and an estimate of costs and payback times. Finally, most utility companies now offer free walk-through audits. (In fact, the larger ones are required to in most states.)

All of these audits have merit, but at best they provide rough guidelines and estimates. "With instrumentation, you measure things instead of just assuming," says Princeton University's Ken Gadsby, who helped develop energy-audit procedures that have been adopted by some commercial auditing firms. "It's a systems analysis. Instead of having an insulation installer do one part, a storm-window salesman do another, and a furnace man another, a professional auditor does it all at once."

## PRESSURE TEST

It's the instrumentation that really sets the pros apart from other audit services. The heart of the system is the infrared scanner and a device called a blower door, a high-powered fan that mounts in an exterior doorway to pull air out of a house. The scanner then detects cold outside air rushing into the depressurized house through leaks in the building skin.

Used alone, the scanner can't find as many leaks. "On the windward side of a building, the scanner by itself works fine," Gadsby explains. "But on the leeward side, it won't see anything. Yet what comes in must be going out. The blower makes all the leaks unidirectional. It amplifies everything and makes it much easier to do the job right."

The blower not only helps find leaks; it enables you to measure them and your success in elminating them. "It gives an exact measurement of the pressure differential between indoors and out-doors," says Bill Munsell, an auditor for a house-doctor service, Energy Clinic, with

**Infrared scanner** spots insulation bypasses and air-leakage sites better than any other tool. It requires temperature difference between indoors and out.

branches in Connecticut and California. "After we do our sealing, we check the pressure difference again. With the right formula, we can tell just how much we've reduced the rate of air infiltration."

Energy Clinic's service generally cuts infiltration by 25 to 30 percent, but the improvement can go as high as 45 percent in a really leaky house.

Locating air leaks is only one aspect of a good audit. Just as important is finding what Gadsby calls "bypasses," routes by which escaping heat can get around your insulation. Bypasses are common in attics and basements, where insulation may be inadequate, poorly installed, or completely missing. They also typically occur around plumbing runs and electrical fixtures. Interior partitions are another problem area, since they're usually leakier than outside walls and they form a chase that runs from the attic to the basement.

Bypasses can also be more subtle. Gadsby gave me an example: "Say you have a soffited ceiling over a shower. There is insulation over the top, but it is stapled at maybe 18-inch intervals, and there is a gap or 'fish mouth' between the staples. The air in the soffit, heated by room temperature, rises and goes out through the fish mouth, and cold air from the attic comes in the same way. You start a little pump going. It's not a direct air leak out of the building, but heat is still being lost."

Besides testing for infiltration and insulation bypasses, a complete energy audit also includes checks of furnace performance, water- and space-heating thermostat calibration, shower-head flow rates, hot-water-pipe and furnace-duct insulation, and more.

A furnace tuneup is particularly important. With a half-hour adjustment, a professional house doctor can improve the efficiency of older furnaces by five to 10 percent. Thermostats are another common energy waster. In Gadsby's experience, it is not unusual to find thermostats that are off by as much as 15 degrees. Bad ones should be recalibrated or, better, replaced with setback types.

## TAKING ACTION

Finding the problems is only part of a good audit. The second part is recommending the right ways to correct them. That's another area where the pro's experience can help. He'll provide you with a printout of suggested actions that can take into account the likely cost-effectiveness of the various options. He will also have marked the leaky spots on the inside of the house (Energy Clinic marks them with pieces of tape) so you or your contractor can start right in with the retrofit work (see box for a list of common retrofit procedures).

**Furnace-testing equipment** checks efficiency of your heating system by measuring draft, soot, stack temperature, and carbon dioxide content of the stack gas.

**Smoke pencil** contains a chemical that combines with air to produce visible precipitate. Where leaks exist, as around this A/C unit, smoke flows to crack.

Some auditors will even do some of the modifications for you. The Energy Clinic is typical. It performs a blower door scan (leaving the treatment to you) for about $175. This is a good alternative if you have already made a lot of fuel-saving improvements or if you want to cut costs by using your own labor. Or you can choose the complete service for approximately $500. Then its auditors will actually plug infiltration sites, caulk and weatherstrip around doors and windows, tune your furnace, calibrate your thermostat, insulate water pipes, and install various energy-saving hardware.

The $500 charge may sound high, but according to Energy Clinic's president Jackson Gouraud, "It doesn't even come close to covering our costs." He says the company offers the service as a loss leader, hoping it will stimulate sales of the wood and coal stoves, Blueray furnaces, and solar hardware that Energy Clinic sells. Besides, says Gouraud, energy audits qualify for the 15 percent federal tax credit, so the cost is really $425. If the modifications save you 15 percent on your fuel bill, for example, the payback could be less than two years—depending, of course, on the price you pay for fuel.

## Here's what your house doctor is likely to prescribe

By itself, an energy audit won't save you any money. Once you get the auditor's recommendations, you must follow through, starting with those that are most cost-effective. Listed here are some of the commonly suggested jobs, though not all are necessarily appropriate for every house. Most of them can be tackled by the average homeowner.

Caulk in places where leaks were found, usually at wall intersections, above moldings, and around window casings.

Weatherstrip around all doors and windows.

Install foam gaskets behind electric-box cover plates.

Seal around pipe and wiring runs.

Install plastic vapor barrier in dropped ceilings.

Weatherstrip and insulate attic hatch.

Seal gaps between foundation and sills.

Tune furnace (this calls for a pro).

Install storm windows or new multiple glaze windows.

Insulate water heater and first 10 feet of hot-water pipe. Turn down thermostat to 120 degrees.

Install low-flow shower heads if rates exceed three gallons per minute.

Upgrade your masonry fireplace with add-ons or convert to a wood stove.

Increase attic insulation.

Install insulating shades or thermal shutters.

# Listening for air leaks: How to spot infiltration with your ears

*Simple acoustical tools and background noise can stem heat and A/C losses*

## By Paul Bolon

Thirty to 50 percent of a typical home's heat literally leaks away—expensively heated air escapes through tiny cracks and crevices, and cold air seeps in. Most of us have taken caulking gun and weatherstripping in hand to perform the tedious chore of trying to stem those leaks. But tiny pinholes and cracks are difficult to spot, and sometimes they can't be seen at all.

Even skilled professionals often believe that they can see most air leaks. But obscured joints, such as a soleplate overlapped by siding, can't be inspected by eye. And even if a hole is found and caulked, there hasn't been a way to check whether the leak has been completely closed. But now there's a new method, based on acoustics, to locate those leaks and make sure they're sealed.

David Keast, an engineer in Cambridge, Massachusetts, investigated acoustical leak detection, working under a contract with Brookhaven National Laboratory. He came up with this surprising result: Most leaks can be pinpointed with the use of simple, everyday equipment.

To see how practical the method is—or whether it works at all—I armed myself with a variety of detection equipment and did my own testing. My results confirmed Keast's: Acoustical leak detection does effectively locate significant leaks, and most homeowners can perform this simple procedure themselves.

Acoustical leak detection relies on a simple physical phenomenon. An uninterrupted path of air, even a very small one, will carry sound through a wall with a volume twice as great as that which the wall itself transmits. To apply this principle to locating air infiltration, you need only a source of sound—noise, really—and a listening tool precise enough to detect point sources of sound. Leaks can be detected by moving a listening tool such as a mechanic's stethoscope

183

along a joint where a leak is likely— around doors, windows, and soleplates, for example—and noting where the sound is louder. Where there is no leak, you hear only sound that the wall is transmitting. Near a hole, however, the sound increases perceptibly.

You place your sound source inside the house or building on the same floor that you are going to check. Good noise sources include vacuum cleaners, dishwashers, washing machines, or, possibly, records or tapes on a stereo or hi-fi system. The sound need only be fairly steady in volume, and broad band—that is, with many different pitches mixed in. Most musical recordings have too much variation in volume to be useful. Perhaps the easiest way to make use of an appliance for sound is to record it on tape, then play the tape on a stereo system at increased volume.

A number of simple tools qualify as good listening devices: a mechanic's stethoscope; a microphone coupled with a headset; a small, thick-wall rubber hose; a hollow-tube headset of the kind airlines supply for listening to music; or a sound meter.

Keast's testing included all of the devices a homeowner might resort to—as well as a battery-powered microphone and earphone headset that is no longer on the market. I used several sound meters and airline-type, hollow-tube earphones in my tests. I played a tape on small stereo speakers for my source of sound. With all doors and windows shut, I first examined the house with just the earphones, chalk-marking places that were noticeably louder. Then I checked my work with a sensitive sound meter.

Using the earphones, it wasn't difficult to distinguish the louder volume resulting from a medium-to-large leak, and the sound meter verified the location of the leaks I heard. After caulking the holes,

**Detecting air leaks** with a microphone and headset is done with a loud source of sound inside. Even small holes transmit about twice as much sound as the wall itself. Graph below gives locations of heat and A/C losses by infiltration. Five to 20 percent of A/C output is lost via leaks. Piling on insulation will not affect leaking.

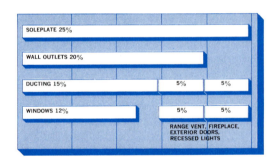

| | | | | |
|---|---|---|---|---|
| SOLEPLATE 25% | | | | |
| WALL OUTLETS 20% | | | | |
| DUCTING 15% | | | 5% | 5% |
| WINDOWS 12% | | | 5% | 5% |
| | | | RANGE VENT, FIREPLACE, EXTERIOR DOORS, RECESSED LIGHTS | |

the sound diminished, confirming the presence of the original leak.

I found acoustical detecting especially effective in locating leaks at soleplates and around doors. (Incidentally, soleplates should be checked once with the sound source on the first floor and again with it in the basement.) Testing was easier with a good sound meter, since it gave a verifiable reading on its scale. Unfortunately, the less expensive sound meters I tested, costing about $50, have scales that begin at 60 or 70 decibels—not sensitive enough to work conveniently.

The only drawback I found in using the simpler listening device was that my ears got tired after a while. Rest periods between listening intervals helped quite a bit. It also helped to use a headphone set to isolate myself from background noise.

Sound along outside walls is louder where solid surfaces meet inside, as at corners and at floor and ceiling levels. It sounds a third louder at floor level than on the wall above, and about two thirds louder at corners. But since it is the relative volume along each single joint that is compared, this absolute difference poses no problem.

Keast originally thought that home centers or lumberyards might rent battery-powered microphones and headsets to stimulate sales of caulking and weather-stripping. But the idea hasn't caught on, and the nifty battery-powered equipment isn't available now. So we're left with the simpler tools.

One final note: Acoustical leak detection does not spot long and complex leakage paths through walls. For this reason, the method is far less useful on houses

**Checking for air leaks** can be done with hollow-tube earphones (inset). When using simple tools like this, it's best to pick a day when it's quiet outside.

already tightly sealed. And it won't tell you whether you have insulation in your stud spaces—or how much.

# Smoke out air leaks

*A high-speed fan and an incense stick can show where heat is escaping your house*

## By Paul Bolon

It's only when the temperature plunges and the oil bill soars that most of us act to make our homes weathertight. The obvious first step seems to be to add insulation in the attics. But far more heat is lost from most homes via air infiltration—air leaks through small cracks and crevices—than through roofs or walls.

In a typical home built five or more years ago, about 30 to 40 percent of the heat produced by the furnace in winter is wasted: warm, humid air escapes outside and cold air seeps in to replace it. Piling on more insulation in the attic or walls doesn't halt the exodus of air, or even slow it down. Most leaks are not through insulated areas.

Although finding and plugging air leaks may seem a small, nagging chore—not one that could give the dramatic results expected from a thick layer of insulation or a set of storm doors and windows—a conscientious job of stopping air infiltration can in fact reduce the heating bill for most houses 10 to 15 percent, and materials for sealing leaks cost little. A well-sealed house also maintains a higher, more comfortable humidity during winter—all the more reason to give reducing infiltration top priority—the step you take before tackling the expensive energy-saving jobs.

How do you find small leaks? There are two standard techniques: (1) By making a visual check for cracks. (2) By feel—holding a hand up, for example, to a window to feel for a cold draft on a windy day. Feeling for cold is not an accurate way to spot leaks—or to tell that they are plugged after you have tried to seal one. Although you can spot some potential sources of leaks with a visual check—an obvious crack under a door, a sill plate in the basement that's not flat on the foundation, or a loose window—you can't see the small air currents that make up much of air infiltration losses.

There is a new method that will do a nifty job of finding those leaks—and you don't have to wait for a winter gale to use it. Observing the direction of small wisps of smoke from a punk, an incense stick or even a cigar that is held near a possible leak is not an entirely new idea, but it's never attained widespread use. However, when that method is combined with some technical innovations, the result is a simple, sure-fire technique.

Engineers have long used air pressurization to measure leakage in buildings (or pipes or other large volumes). That practice requires a very powerful fan and accurate instruments to monitor the inside pressure and the volume of air that the fan is moving. These measurements are then used to estimate the usual flow of air out of the structure.

It was recently found that, in the case of a house, a powerful high-speed fan will work just as well as the sophisticated equipment used by engineers; here, however, the aim is not to calculate the leakage rate, but to produce enough pressure to maximize the air flow out of any leaks. Then, using a small source of smoke, leaks are spotted by noting where the smoke is drawn outside.

It was with some skepticism that I tested this new system. But my results were good—even a little startling.

I began by mounting a large high-speed fan in a window. I cut a piece of ¼-inch plywood in a rectangle as tall as the fan and just wide enough to fill the remaining space in the window jamb. To secure the setup, I lowered the bottom sash onto it. Cracks around the fan and plywood were sealed with tape. (Heavy polyethylene, secured well with masking tape, could be used instead of the plywood.)

The fan must, of course, be set to blow air *into* your house. Also, all outside doors and windows must be closed and all interior doors opened. If you have a fireplace or a wood stove, you must close the damper or use paper or some other material to block the flue. (Don't forget to remove the paper after your test is finished.) Switch off the power to your furnace while testing—don't just turn the thermostat down. Seal off the basement from the rest of the house and test it separately. All this sealing off is done so the fan can build as much pressure as possible inside

## Where to look for air leaks

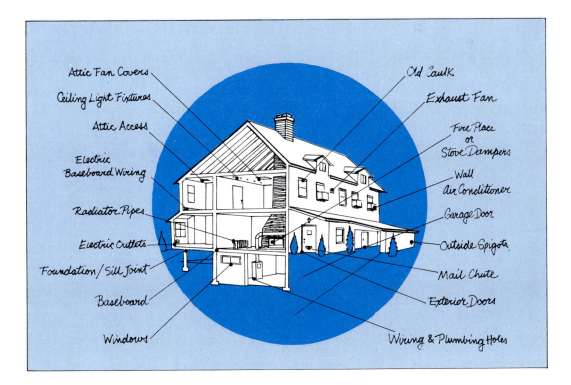

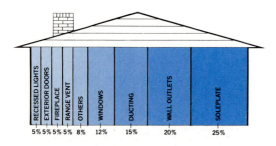

## Air leak culprits

country to light firecrackers or to keep bugs away from the patio on summer nights—does as well.) I started with a window I was pretty sure would show some leaks. There weren't any along the side: As I moved the stick up and down near the sash/window-stop joint, the smoke rose in an uninterrupted stream—if air had been rushing out at any point, smoke should have been drawn into the joint. Next, I trailed the stick along the joint between the sashes. The column of smoke quickly disappeared. Then it reappeared—outside the window! The leak was obvious with the smoke. I could not have been sure that there was one just by looking.

the house. More air rushing out the leaks pulls more smoke with it, making leaks easier to detect.

To begin my leak hunt, I turned the fan on and lighted a stick of incense. (Punk—a smoldering stick sold in some areas of the

I continued testing along window joints, doors, baseboards, electric out-

## Overhead view of closed room

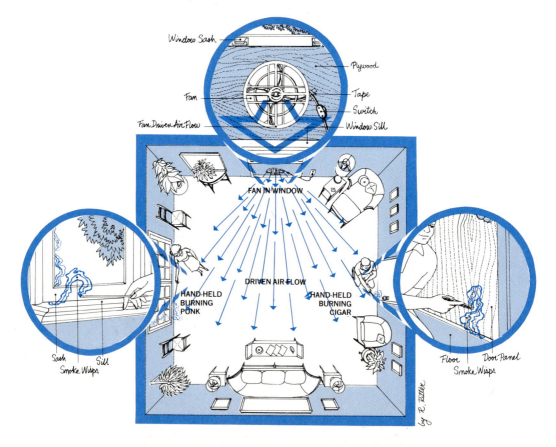

## Another way to track down air leaks

You would think that any thermometer could pick up areas of infiltration. Not so. It would have to be one that is both very sensitive and immediately responsive.

And that's the idea behind the Heat Sleuth—basically, it's a quick-reacting, electronic thermometer—and I found it quite capable of detecting air leaks. The 3 × 4-inch, 2½-inch deep primary box has a two-inch meter face, a knob and an on/off switch. Attached to this with a four-foot cable is the "wand," a probe you pass along the area being tested.

Using the instrument is simple enough: (1) Turn off your furnace and wait until the fan stops and stray air currents are eliminated. (2) Flip on the Heat Sleuth switch and turn the knob until the needle rests at the midpoint zone labeled "Normal Temp." (3) Pass the probe along potential leakage points (around windows, baseboards, ceiling moldings, etc.). When you hit a spot where cold air is infiltrating, the needle drops into a zone labeled "Probable Heat Leak"; in summer, when air conditioning is in use, infiltration will cause the needle to rise into a zone marked "Probable Hot Spot." After you've eliminated a leak, a second check with the meter tells you how effective your work was.

For detecting heat loss, the device works best on a cold, windy day with a differential of 30 degrees or more between indoor and outdoor temperature; at less differential, the meter is less sensitive, since there is less infiltration. The same principle applies in reverse for heat gain in summer. The Heat Sleuth can also be used outdoors on a still day, but temperature variations caused by slight breezes or solar exposure tend to make readings erratic.

This is a sensitive instrument. Variations of only ¼-degree can cause a 1/16-inch movement of the needle. This, coupled with the fast reaction time, make it useful for a number of other purposes. (One I've already discovered is testing the gasket seal on refrigerators and freezers.) This flexibility, plus potential savings in heating and cooling costs, could pay back the price in a short time. The Heat Sleuth is made by Enertron Corp., 241 Crescent St., Waltham, Mass. 02154. Phone: (617) 893-3532.

lets—all the places (see drawing) that are potential sources of leaks. As I worked my way around the house, I chalk-marked each leak. I found the largest ones around doors and in the basement at the joint where the sill plate rests on the concrete foundation wall. The smoke telltales were adept at pinpointing small leaks, too.

After testing all possible sources of leaks, I went to the hardware store for the caulking, weatherstripping and threshold seal I would need to fix them. As I applied weatherstripping, I worked with the smoke stick. You can rely on the smoke stick to make sure you've done a thorough job. However, you can't use it where air currents from the fan blow directly on the spot you're working. You have to move the fan at least once to check the window in which it was first mounted. If your house is really big, you may need to mount two fans at once in order to get enough air pressure inside to test easily.

If you examine your house carefully using the smoke detection method, and act on your findings, experts say you should be able to cut your rate of air infiltration in half, especially if you don't already have storm windows. (Stormers are probably the easiest way to help seal doors and windows.) This typically translates into a 10 percent or more reduction in your heating bill. Stopping leaks also reduces the cost of air conditioning in the summer, but not as much.

If your house always seems stuffy or full of stale air, don't bother searching for leaks. That still air means it may already be pretty tight. The tiny leaks in our homes are how we vent odors and gases from plywood and concrete, stove gases, tobacco smoke and other household pollu-tants. Such pollutants are present only in small quantities, but they have to be removed by slow, constant changing of house air. Unless your home was carefully built for energy efficiency in the last few years (in which case you may need a heat exchanger or ventilation fan), you can plug all the leaks you can possibly find and enough tiny pinholes will remain to provide a healthy indoor atmosphere.

Also note that your furnace needs a source of outside air if you burn gas or oil. It's much better to run a vent directly from the outside to the burner than to draw cold air from big leaks—that just keeps your entire basement cold and wastes heat.

# Wonder windows— Four new-technology glazings that save energy

*Two let in more sun; two keep in more heat*

By V. Elaine Smay

Windows are rather like the little girl with the little curl in the middle of her forehead: When they are good, they are very, very good—they let in the sun to illuminate the interior and warm your house. They can provide a nice view, too.

But when they are bad, they are horrid—probably at their most mischievous on a cold winter night, when they may let out much more heat than they collected during the day. They're not so nice either when it's 99 degrees and the sun is streaming in.

Multiple glazing improves their thermal behavior. But a double-glazed window still transfers heat six times as readily as a moderately well-insulated (R-11) wall. And even a triple-glazed window passes heat four times as readily. "About five percent of our national energy consumption goes to offset heat loss through windows," says Stephen E. Selkowitz, who heads a Department of Energy-sponsored program at Lawrence Berkeley Laboratory aimed at improving windows' energy efficiency. "That's about 1.7 million barrels of oil a day." And it may account for 12 percent of your household energy use.

No one has yet come up with a window that's a real Pollyanna, giving all possible benefits all the time. But four new products (three available now and one coming soon) do offer significant improvement—and by opposite means. Two use an ingenious transparent coating that reflects heat. One of these is a coated glass, called Low-E glass, made by Airco Temescal of Berkeley, California, and Guardian Industries of Carleton, Michigan. The other is a coated plastic film, Heat Mirror, made by the Southwall Corporation in Palo Alto, California.

A third new product, made by the 3M Company, is also a thin plastic film. This one, called SunGain, lets in more solar

191

**Transparent-plastic film,** SunGain from 3M, lets in more solar heat and light than conventional window glass. One or two layers of the film are put in the air space between two panes of glass to make triple- and quadruple-glazed windows.

heat than glass. The fourth product, glass with a reduced iron content, does the same. Low-iron glass, Solakleer, is made by General Glass International, New Rochelle, New York.

All could help make energy winners of more windows and make passive-solar design easier to live with by eliminating the expense and nuisance of movable insulation to cover glass.

The coating on Heat Mirror film and Low-E glass is similar to what has been used on airplane windshields since World War II. It is electrically conductive, and sending a current through it de-ices the windshield. But such glass is costly and is produced only in small quantities.

"In 1969," reports Day Chahroudi, a physicist then with Zomeworks, "I learned of the windshield technology and decided it could be adapted for glazing systems to create a transparent insulation." In 1974, Chahroudi joined the architecture department of the Massachusetts

Institute of Technology. There he, John Brookes, and Sean Wellesley-Miller developed the technology. The Southwall Corporation was later formed by the group.

Airco Temescal makes vacuum-deposition machinery for the semiconductor industry and got interested in the coating because it is applied with similar machines.

Southwall's Heat Mirror, a two-mil polyester film, is used in triple-glazed windows with one layer of film and two panes of glass (see diagram). A number of window makers now sell Heat Mirror windows and more should be soon. Low-E glass is used in double-glazed units. (E stands for emissivity, a measure of a material's infrared-reflectance characteristics.) Guardian Industries markets Low-E windows.

The coating on these glazings reflects long-wave infrared radiation (radiant heat), the kind emitted by room-temperature objects and people (see box, page 194). Unlike solar-control films, it transmits most of the solar spectrum. Solar-control films darken the window from the inside and give it a mirrorlike look from the outside. With Heat Mirror film, the window isn't visibly different from triple-glass windows. Heat Mirror lets through 72 percent of the solar spectrum and 88 percent of visible light. Low-E glass has a slight bluish cast from the outside, but looks clear from inside. It transmits 71 percent of visible light and 56 percent of the solar spectrum.

Windows with a Heat Mirror film or Low-E glass will have the same convective and conductive heat loss as do those with equal panes of conventional glass and equal air spaces. But they reduce heat loss through radiation. That can amount to ½ to ⅔ of the total heat loss. Glass absorbs radiant heat and reradiates it in all directions—outside and inside. These coatings reflect 85 percent of the radiant heat, so most stays in the room.

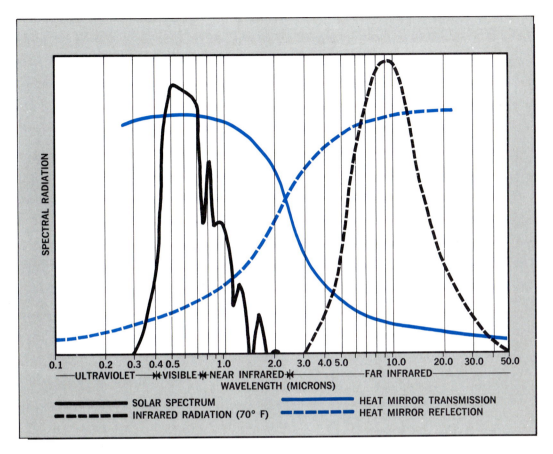

**Heat Mirror film** is engineered to transmit solar radiation in the visible and near-infrared portion of the spectrum (solid blue line). But it reflects the far infrared (dotted blue line), the long-wave radiation emitted by room-temperature objects.

The net effect can be seen by comparing U-values (a measure of heat flow in Btu/hour/square foot/degree F). The lower the U-value, the less readily the material transfers heat.

A single-glazed window with standard glass has a winter U-value of 1.13. A double-glazed window with a ½-in. air space is 0.56. A triple-glass unit with two ½-in. air spaces is 0.34. A window with one Heat Mirror film, two panes of glass, and two ½-in. air spaces has a U-value of 0.23. "That's 32 percent lower than triple glass," says Alexander Tennant, Southwall's vice-president for marketing, "and you reduce the weight 33 percent."

Airco-Guardian's Low-E glass in a double window has a U-value of 0.33, about equal to triple glass but with only the bulk and weight of double.

The energy advantage of such windows should be even greater than these numbers suggest. Window surfaces will be warmer, so you won't radiate as much body heat to them and drafts will be reduced. And you can raise humidity

## How the coatings work

Heat Mirror film and Low-E glass are coated with a three-layer sandwich only a few hundred atoms thick. The center layer is a conductive metal—gold, silver, copper, or aluminum. The metal layer is between two layers of a metal oxide, a dielectric or insulator. The metal layer reflects long-wave infrared energy as the free electrons in the metal (those unattached to an atom) interact with the electric field in the IR. It also reflects some of the desirable solar radiation. "But the higher the frequency of the radiation, the less efficient the reflection," explains Dr. Stephen Meyer, a physicist with Southwall, "so even if the metal layer were used alone, about 50 to 60 percent of the solar spectrum would get through." But the two metal oxide layers improve the solar transmission to around 72 percent. "As the higher-frequency wavelengths hit the metal, there's a 90-degree phase jump," Meyer explains. "That's what causes most of the reflection." The metal oxide layers are engineered to cancel that jump, and thus let the solar radiation pass through the metal. "It's similar to impedance match-ing in electrical transmission," he adds, "where you use a transmission line to transform an impedance and bring the voltage and current in phase with each other."

The coating on 3M's SunGain film is an ultrathin layer of a transparent metal oxide. Aluminum, magnesium, or zinc oxides can be used.

As light passing through one transparent medium (air, for example) hits another with a different index of refraction (glass or polyester film), part of it is reflected. The greater the difference in the two indices of refraction, the greater the reflection. SunGain film reduces that reflection with a surface microstructure. "We call it a graded index of refraction," says George Ruth, 3M's project leader for SunGain's development. The surface, magnified 100,000 times, looks rather like velvet fabric—tiny needles give it a "nap." These needles, roughly cone-shaped, gradually increase the metal oxide's effective index of refraction from tip to base. Thus it acts as a transition between the air and the polyester film.

without getting condensate on the windows. These factors make you feel warmer at lower thermostat settings. "The net benefit can be a three to five percent additional improvement," says Tennant. In summer, you'd have less heat gain.

In a climate where heating costs more than cooling, it is not just heat loss that counts, however; it's the net energy transfer—the amount of solar heat the window lets in minus the room heat it lets out. Heat Mirror film and Low-E glass let in somewhat less solar heat than conventional window glass. High-transmission glazings—3M's SunGain and General Glass International's Solakleer—let in

more. SunGain transmits 93 to 96 percent of the incident solar energy and Solakleer transmits 90 percent. Standard glass transmits 84 percent.

SunGain is a polyester film with a coating that reduces surface reflectance (see box). Weather Shield, a window maker in Medford, Wisconsin, is using SunGain film as inner glazing between the two panes of glass (see diagram). The company offers triple- and quadruple-glazed windows, doors, and patio doors. The quad-glazed units have a U-value of 0.26—about the same as triple-glazed windows with one pane of Heat Mirror film and with a comparable benefit from warm window surfaces.

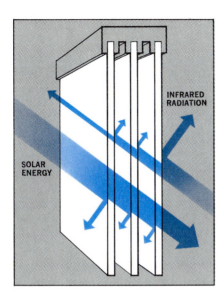

**Triple-glass window** transmits most of the incident solar radiation, but each pane reflects and absorbs some. Infrared radiation from room objects is absorbed by each pane and reradiated both ways.

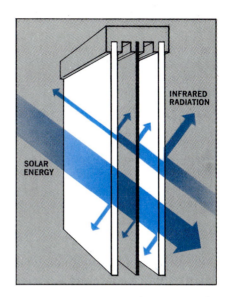

**SunGain film,** the inner glazing here, transmits more solar radiation than glass. It does not block infrared transfer quite so well, but its thinness allows a larger air gap so the total U-value is equal.

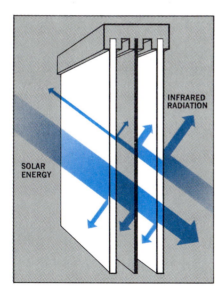

**Heat Mirror film,** the inner glazing here, admits less solar radiation than glass. But it saves heat better by reflecting back most of the infrared radiation emitted by the interior pane of glass.

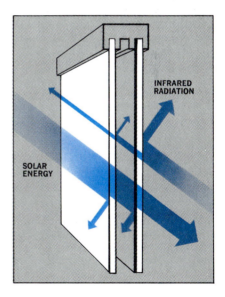

**Low-E glass** transmits less solar energy than Heat Mirror film. But the window has fewer panes to reflect and absorb it, so total solar gain is a bit better. Low-E's IR reflectance equals Heat Mirror's.

Low-iron glass improves solar transmission by reducing its absorption within the glass, which is caused by impurities in the glass, mainly iron oxides. Solakleer glass has fewer impurities. It is sold to window manufacturers who put it in single-, double-, and triple-glazed windows.

Adding a layer of Solakleer glass or SunGain film lowers a window's U-value as much as adding a pane of conventional glass. But since you don't sacrifice as much solar heat, the net energy gain is improved. The 3M Company calculated the net energy gain for south-facing windows in a moderate climate (Indianapolis, Indiana), comparing quadruple-glazed SunGain windows with standard double glass and night insulation (R-5) in place 14 hours a day. The SunGain windows came out slightly better.

But all these new technologies cost more than standard windows. Airco Temescal expects double-glazed windows with one pane of Low-E glass to cost 12 to 22 percent more than conventional double-glass windows, but a little less than triple glass. Hurd Millwork, Medford, Wisconsin, will sell woodframe triple-glazed windows with one pane of Heat Mirror film for seven to 10 percent more than triple-glass windows. Weather Shield's wood units range from eight to 12 percent more for triple glazing with SunGain film than for triple glass; quad glazing (two panes of SunGain) is about 17 percent more than triple-glass units. Solakleer costs about 20 to 30 percent more than standard window glass.

Would you soon make up in energy savings for these higher initial costs? "In the long run," says Selkowitz, "I think all of these products will be cost-effective, but the specifics depend on fuel prices, climate, the orientation of your windows, and their cost."

Which type would be best for your windows? "In general," Selkowitz explains, "the colder and more overcast the climate, the more important the U-value and the less important the solar transmission. In a climate that's sunny but needs significant heat, you should pay attention to solar transmission. Furthermore, on the north side of the house a low U-value is more important. On the south, you might want high solar transmission."

Selkowitz offers this additional advice: "The biggest concern about the coatings is how well they will last." All have been extensively tested and seem to hold up very well as long as the seal around the window holds. "But we'll only know for sure," he points out, "after they've been in the field 10 to 15 years."

# The preframed aluminum way to window replacement

According to the Architectural Aluminum Manufacturers Association, there are more than three-quarters of a billion windows in U.S. homes. So how come you got the ones that leak air? Well, you're not alone.

Most homes still have single-pane windows that cost, it's estimated, from $20 to $100 a year in heat loss through conduction and drafts. But you can cut that loss in your home by 50 percent or more, claims the AAMA, with new double- or triple-glazed windows. And window replacement can be simplified.

Preframed aluminum windows are available in a variety of sizes and styles through local lumberyards or home-improvement centers. Basically, you measure the frame, carefully remove the old window, and install a replacement window system of matching size. There are no structural changes, so no carpentry is needed; you may not even need to repaint. And since the new window is installed from inside the house, you won't need scaffolding and the job can be done at any time.

These photos show the simple procedure. The only tools you'll need are hammer, screwdriver, broad chisel, ⅜-in. drill, square, plumb, light hacksaw, and caulking gun. Result: easy-to-operate airtight windows that save money and add to the value of your home.

**Remove molding and inside stops** with a chisel and save for reuse. Remove spring balances, pull out bottom sash (left), and cut weight cords. Pry out parting stops at top and each side of the window frame, as shown at right. This will allow you to slip out the top sash.

**After removing the new sash** for easier handling, place the header expander on top of the frame. Or if the windows have fins, then trim the fin at the proper serration.

**After a trial fit,** remove the frame and caulk against the blind stop at both top and sides.

**Push the window frame** against the caulked blind stops and drive the first screw. Be sure you have the proper screwdriver at hand.

**If the frame has alignment screws,** adjust them until both sides are plumb. If not, add shims at the center of both jambs until square.

**Once the frame is square,** complete the installation with additional mounting screws. Then you place both sashes in their tracks.

**Apply caulking** all around the window framing as an inside seal. This, along with the blind-stop caulking, will prevent any air leaks.

**If an expander header is used,** lift it to the top of the opening. Drill mounting holes and insert wood screws to anchor the header.

**Replace the inside stops and molding,** which were previously removed. Use care with the hammer—or you may not yet be finished.

# Self-storing storm window

By Paul Bolon

Storm windows cut winter heat loss through single-pane windows by about half. But well-made aluminum sash-and-screen types are expensive, with payback periods of three to four years—or longer. That's a little discouraging since, after all, we're trying to save money spent on fuel. The only way to cut the cost of storm windows was to use single-pane stormers, which had to be remounted in the window each winter. But now there's a new system that may offer a cheaper alternative.

The Wind-O-Way "storm window" is basically a roller shade with Du Pont's Mylar film in place of the usual vinyl material. The bottom of the film is attached to a plastic bar for lowering and raising, and the bar and film are sealed along the edges and bottom of the window with extruded plastic runners. Wind-O-Way's roller mounts like a typical shade's and is neatly hidden by a cover. All the parts are made from the same white or colored plastic, which produces an uncluttered, clean-looking window (photo).

The Wind-O-Way is being sold as a kit, which homeowners cut and trim to fit each window. A kit that fits windows up to 48 by 60 in. tall costs about $26. A smaller and a larger kit are available, too.

Tinted and reflective Mylars are optional and can be very useful on south-facing windows in summer. For more information: MARC Inc., 228 Park Avenue W., Mansfield, Ohio 44902.

**Cross sections of** Wind-O-Way's vertical runners and header cap (with roller support) are in photo, bottom. Parts can be mounted using pressure-sensitive backing, brads, or screws.

# Super-tight storm windows

## By Ray Hill

Ideally, an old, leaking window should be replaced with a modern double- or triple-glazed unit. When that isn't possible, installing a storm window may have to suffice. A problem with many storm windows, though, is that their pile weatherstripping can let in air.

Several storm windows now on the market significantly reduce that air leakage with a new type of weather-stripping—Fin Seal. Like conventional storm-window weatherstripping, Fin Seal uses pile to keep the windows from rattling. But a flexible fin in the middle of the pile (see drawing, lower right) blocks out the air that might otherwise leak through.

Its developer, Schlegel Corporation, of Rochester, New York, says that the average storm window leaks two cubic feet of air per minute for each foot of crack perimeter. Testing, Schlegel says, has shown that windows using Fin Seal leak approximately one-fourth that amount.

To determine whether a storm window uses Fin Seal weatherstripping, check for the flexible fin in the middle of the pile.

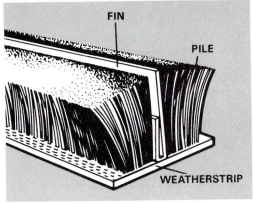

# Windows tight? Now maybe it's time for energy-saving insulating shades

*R-values? Edge seals? Here's how to evaluate the claims before you buy*

## By A. J. Hand

Today, window shades are expected to do a lot more than darken the bedroom when you want to sleep late. Properly installed, even an ordinary shade from the dime store can block a lot of heat loss. A specially designed thermal shade can truly insulate your windows.

Just a few years ago, there were no such things as thermal shades. Today, there are a dozen or so, with more scheduled to appear on the market.

Thermal shades are not for everyone. To help you decide if they're for you, we offer a number of points to look for.

Thermal shades come in a variety of versions, but all work on the principle that air is a good insulator. A tight-fitting shade creates an insulating air pocket that prevents warm room air from being drawn to and chilled by a cold window surface.

Most thermal shades also use some kind of heat-reflective material—usually metallized plastic film, the stuff used to reject summer heat.

Makers of translucent shades such as the N-R-G shade and the Insealshaid system say metallized shades bounce infrared rays back into the room. (Other shade makers cover the metallized plastic with a decorative fabric or finish.) This infrared reflection property may conserve some extra heat—though not much unless there's a radiant heater in the room. Here's one advantage of these translucent shades, however: On cloudy winter days you can pull the shades for insulation, yet still see out.

On the theory that more is better, most of the shades have multiple air-trapping layers. Shades such as Curtin Wall and Independence I—formerly the IS High "R" shade—have four layers of metallized plastic that offer multiple air pockets.

An alternative approach to window

**Glittering gold** Curtin Wall has multiple layers that inflate when shade is pulled down. Don't want the gold lame look? Manufacturer will attach buyer-supplied fabric to front of shade with Velcro so frabric can be removed for cleaning.

insulation is to cover the window as you would a bed—with a thick, padded blanket such as Window Quilt.

Finally, some shades do double duty as solar collectors. Both Thermo-Shade and Curtin Wall have automatic controls that raise the shade when the sun is shining and lower it at sunset or when clouds cut off the heat. Insealshaid features a thermally controlled vent system that alternately closes and opens to trap, then release, warmed air to the room.

Thermal shades don't come cheap. But if you are dismayed at the prices, note the Sol-R-Shade listings. This do-it-yourself system can be a money saver. Components include a variety of shade materials such as reflective films, thin foams, and laminations combining fabric and film. The iron-on edge seals have self-adhesive window tracks. Clip-on bottom battens have foam strips to seal the shade against the sill. Spacers between battens hold the shade plies apart to create insulating air spaces. And the shade roller is a simple cord-operated system. All the hardware for the Sol-R-Shade is presently being imported from Japan, but there are plans to produce it in the U.S.

Who should install thermal shades? If you have not yet insulated your roof and walls, caulked and weather-stripped thoughout your house, and installed storm windows, you're not ready for thermal shades. It makes little sense to spend $4 to $10 a square foot for thermal shades when R-19 insulation in your attic costs about 25¢ a foot and can save far more energy.

If you are ready for shades, here are purchasing guidelines.

First check R-value, the rate at which

# How insulating shades seal in the air

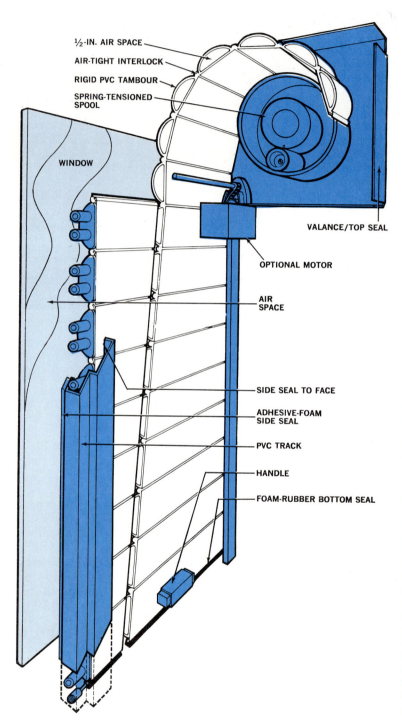

½-IN. AIR SPACE

AIR-TIGHT INTERLOCK

RIGID PVC TAMBOUR

SPRING-TENSIONED SPOOL

WINDOW

VALANCE/TOP SEAL

OPTIONAL MOTOR

AIR SPACE

SIDE SEAL TO FACE

ADHESIVE-FOAM SIDE SEAL

PVC TRACK

HANDLE

FOAM-RUBBER BOTTOM SEAL

**Like the top on a roll-top desk,** Thermo-Shade's tambours pull down smoothly from its storage valance. Though its tambour design is unique, Thermo-Shade is a good example of the design elements needed to make a shade an effective window insulator. All four sides of the shade are sealed. The shade rides in flexible side tracks; foam rubber seals any cracks between shade bottom and window sill; and the valance seals off the top. Thus, a cushion of insulating air is trapped between shade and window. A second air cushion is provided by the hollow tambours. As a bonus, the white PVC has heat-reflecting properties.

the shade blocks heat loss. Surprisingly, you're not looking for the highest possible rating. The R-value of a shade is based on the amount of heat loss it prevents. Obviously the shade does nothing to stop heat loss when it's raised during the day. So if you use a shade normally, it's impossible for any shade to cut daily heat loss through a window by more than about 58 percent.

Thus, the law of diminishing returns sets in quite quickly with window insulators. Shades with a rated R-value of 11 or 12 provide only marginally better insulation than do these with an R-value of five.

The moral? Don't spend extra money just to go beyond R-5. Exception: If you're shopping for a shade that you plan to leave closed constantly during the heating season, an R-value over five can make sense. On the other hand, if you plan to block the window off all winter, you don't need the convenience of a shade. A pop-in shutter can do the job as well for far less money.

Air seals are another key checkpoint. Blocking the flow of air is a vital but tricky job. Without good seals, a shade can't prevent heat loss. Escaping warm air may also cause condensation that can damage window and sill.

But it's also possible to go overboard on expensive seals, just as it is with R-value. The seal's job should be to keep room air away from the window, not to prevent infiltration of outside air into your home. That's a job for caulking and weather-stripping. In practice, seals should be able to keep nighttime condensation down to a level that will quickly evaporate next morning when the shade is opened.

Good seals are especially important on shades that will be left closed during the day as well as at night, because any condensation that does form won't be able to escape.

How can you check for good seals? If you have an actual shade to examine,

press against its face with both forearms. One with good seals will resist.

Unlike the insulation in your walls and attic, an insulating shade must be able to stand up to a certain amount of abuse. Before you buy, examine a shade for possible weak spots. Operate it and check the mechanism. Does it work smoothly, or does it require force that might damage the shade in time? Does the shade rub excessively against its seals? Is it made of durable materials? A rigid vinyl shade might be a better choice than one made of a delicate fabric if you have a home full of active children.

Be sure the shade is easy for you to operate. If a shade is clumsy to work you may fall into the habit of "forgetting" to use it. A shade that isn't drawn at night is a waste of money. Before you buy a shade, try opening and closing it a few times.

Be aware of potential operating problems that might not show up till you get the shade home. For example, a heavy shade on a conventional, spring-loaded roller might work fine on a small store display, but might be difficult to raise and lower if you installed it on a door or an extra-large window at home.

Also, make sure you can live with the appearance of a shade before you buy it. A shade that's too ugly to pull down has an R-value of zero.

What about price? If a thermal shade were nothing more than a piece of insulation your main goal would be to get the best performance at the lowest cost. But since a shade is also a decorative part of your home, only you can decide what is a reasonable price to pay. If shade prices seem high, consider this: Many people spend as much as $5 a square foot for shades or drapes that are purely decorative and offer virtually no thermal benefits at all. If you deduct the money you would normally spend for decorative shades from the price of a thermal shade, it will seem a much better bargain.

# Give your rooms sun control with easy-to-install mini-blinds

*Toss out dusty drapes, end curtain clutter: Mini-blinds do the whole job better*

## By Al Lees

They're a classy update of the Venetian blind, but their arched slats are paper-thin aluminum and only an inch wide. And you can customize them in an array of baked-on colors that make draperies redundant.

Mini-blinds are as functional as they are handsome, offering not only privacy but precise control of ventilation and solar exposure.

Do as I did and order all color bands in duotone slats (color on the front, white on back) and you've got an affective shield against unwanted solar gain this summer.

Pivot the slats to their opposite extreme on winter nights and you'll bounce indoor heat back from wasteful contact with cold glass. And when you *want* solar heat to flood your room, tilt the slats to match the sun angle—or stack them all neatly at the top of the window casing by tugging the lift cord.

The blinds are a cinch to install—either within the casing, as in most of our color shots, or overlapping it. Remember, all these blinds are custom-made and can't be altered once they're assembled. So measure and order carefully—especially for mounting within casings. Blinds come as narrow as 12 inches or as wide as 142 inches. Any width of 55 inches or more should have a center support bracket. For narrower blinds, the two boxlike end brackets suffice. They come with screw holes in three faces so they can be mounted to the face of the frame, within the sides, or to the ceiling. The headrail slips in, and the lids snap shut.

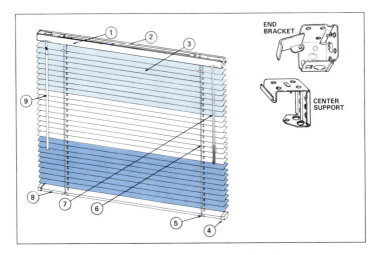

**Outsized and irregularly spaced** windows created awkward wall in author's loft and were single-glazed energy-wasters. Unifying draperies would have cost a fortune and been clumsy to open for view.

PROBLEM

For mini-blind solution, see next page.

SOLUTION

**Identically striped blinds** in six colors unite windows. Graduated effect, with bands narrowing as they darken in tone, is even more pronounced at night. All slats have off-white backs, which bounce out summer solar heat. Blinds by Flexalum.

## LIVING ROOM

Another example of carry-across stripes is designer Ann Heller's room for Levolor with a stair-step of bold colors across an off-white ground. Here, "duotoning" (white backs) wasn't needed.

## STUDIO

Single slats of darker color are matched by window frame to dress up this window without counter-space-wasting curtain. Designer Margot Gunther uses kitchen cabinets for storage.

## KITCHEN/DINING
Super graphics unite kitchen with adjacent dining alcove. Green and plum diamond, wrapped around corner, picks up colors striped across both Levolor blinds. Hang blinds before striping.

## Mini-blinds also solve non-window problems

Pullman kitchen tucks behind floor-length blind that's easily raised for access to work counter. At right, narrow blinds are mounted within room-divider frame to screen off dining area without losing air circulation. Both settings by Flexalum.

# Save energy with ordinary roller shades

## By Paul Bolon

Simple roller shades may be the easiest and most cost-effective way to reduce your winter fuel bill. David Buchanan and Maureen Grasso, researchers at Cornell University, found that ordinary shades cut window heat loss by a surprising 30 percent. Shades with an aluminized surface that reflects radiant heat back inside the house conserve 45 percent of the normal heat loss.

There is only one secret to getting this kind of performance from shades. They have to fit snugly in the window. Shades must be mounted inside the window frame and hang an inch or less out from the bottom sash, with only ¼-inch clearance on each side. When unrolled, they must touch the sill and leave less than an inch gap above the roller. A shade installed more than an inch from a window or that does not close the air gaps is dramatically less effective.

Buchanan and Grasso achieved their results testing off-the-shelf, vinyl-coated products. They made their own reflective shades by adding Du Pont's Mylar film to a regular shade.

How important are windows in conserving heat? And how do shades compare in cost and savings with other heat-saving solutions for windows? In a house with 5½ inches of fiberglass insulation in the ceiling and 3½ inches in the walls, about 25 percent of the total heat loss is through the windows. Only attics and ceilings lose more. Glass is such a good conductor of heat that even small windows create strong convection currents when it's cold outside. These air currents increase the chilling effect of glass by bringing more air into contact with the window and circulating the cold air throughout the room. Close-fitting shades isolate this circulation from the room and minimize the volume of air that the window cools.

To test the effect of completely blocking the windows (for best possible heat savings), Buchanan and Grasso added simple wood channels on the sides and sealed the air flow above the roller with denim. This arrangement increased the heat savings with regular shades to 40 percent and with Mylar shades to 55 percent—which about equals the performance of storm windows. Using channels also counteracts the tendency of shade edges to curl.

## How to construct side tracks

To make side tracks or channels for a shade, first examine your windows. If the window stop, or guide, is less than an inch wide, simply add a small piece of wood—about ⅜ by ⅜ inches will do—and use the stop as one of the tracks. Leave about ⁵⁄₁₆ inches between the stop and the wood for the shade to run in. If the stop is wider than one inch, you can mount the shade brackets on it and use two small pieces of wood or a single piece of wood with a groove, for the channels, as shown in the figure. You can also replace the stop with a track arrangement. The track should extend from two inches below the roller to the sill. The top seal is tacked or stapled above the roller, and can be almost any flexible material: extra shade stock, plastic or heavy cloth. It must be heavy enough not to be affected by rolling the shade. The shade may be closer than an inch from the window—just so it doesn't touch the glass. Measure for shades after you've made the channels, making certain to leave ¼ inch clearance on both sides. For a perfectly fitted shade, screw in the first bracket, unroll the shade in the window, then mark the placement for the second bracket, and screw it in.

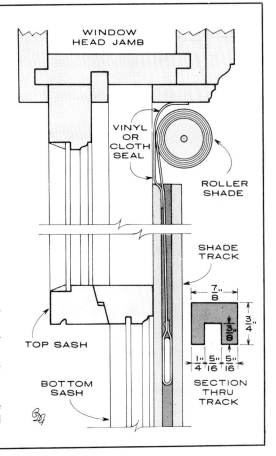

## COMPARE THE ALTERNATIVES

- Storm windows reduce heat loss through windows by about half. They are a long-lasting investment and are relatively expensive. Their payback period in fuel savings will be at least three years—more likely four or five.

- Fabric does a wonderful job of keeping your body warm, but when used for curtains it does almost nothing to insulate windows. If curtains are lined with a vinyl or other impermeable liner, they partially reduce the flow of air. But conventional installations do not effectively seal air flow around a cold window.

- Venetian blinds also have little value as a barrier to the cold.

- There are some insulating shades that have several layers of film and Mylar running in channels two to four inches wide in the window frame. These may require considerable modification of the window frame to install, and they are relatively expensive. Also, they are best suited for large areas of glass, such as picture windows and sliding patio doors, which are commonly of single-pane construction.

Roller shades are cheap ($6 to $10), install easily, and have a payback period of one or two years, even without channels. They are ideal for cutting heat loss in little-used rooms. Drawbacks are the occasionally balky roller springs in older shades and the need to adjust shades manually. If you have shades already mounted on the window casing, it's easy to cut them down yourself to fit the inside dimension.

Shades will save about the same percentage of heat loss if you already have storm windows, but they will be less cost-effective because the storm windows will have already halved the loss. If, for instance, storm windows and tracked Mylar shades are both installed, 75 percent of the heat lost through single-pane windows will be saved.

During the day, you can gain heat by raising the shades in sunny windows. And if you air-condition your home, opaque white shades will save more on your energy bills in summer than in winter.

# Insulated garage door plugs a major heat leak

## By Evan Powell

Even if you lock your garage door, your attached house could still get robbed—of valuable heat. Most such doors are an open invitation to drafts and winter cold. My old door was no exception, so I installed an insulated garage door recently introduced by J. C. Penney.

The polyethylene, aluminum-frame door is made with tough inner and outer panels. While mine uses an air barrier between the panels, others are available with a polystyrene foam-core insulation.

Installation is more critical than with a conventional door, since each panel must interlock to prevent air flow. The photos show some of the major steps and a few tips learned in the process.

A heavy-duty seal is included that attaches to the bottom panel. That should be sufficient for floor contact, but you'll have to provide your own seals for the sides and top. I used a large, hollow-core window seal from Stanley. When you're done, the polyethylene panels can be painted with exterior latex enamel if you don't care for standard white.

During the past winter, I watched the door actually pulsate from night winds.

But when I passed my hand across and around the inside of the door, I found it kept infiltration to a minimum. While its R-value can't approach that of a four-inch insulated wall, it's certainly far superior to a conventional panel garage door. Price: $170 to $610, depending on door width (8 to 18 ft.).

**Large, hollow-core weatherstripping** must be added at the top and sides of the door.

**All roller brackets** (there are four different types) should be temporarily laid on each panel before permanent assembly. Use a thread-locking compound on bolts.

**Bottom panel** is installed first, leveled, and temporarily nailed in place. Successive panels are stacked on top until door is complete, and then fastened securely.

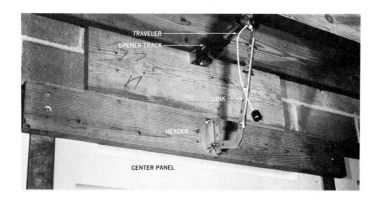

**With an automatic door opener,** a wood or metal header must be installed across the upper panel to spread stress equally throughout the aluminum frame.

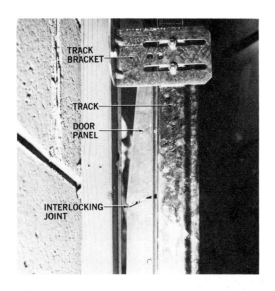

**With all panels in place,** rollers are positioned in brackets, track is installed, and nails are removed. Note how panels interlock, as shown in side view.

# Stop heat leaks through basement walls

## By Evan Powell

Basement walls are an important area of heat loss that is often overlooked. To insulate a foundation wall: dig a trench six inches to two feet below grade, depending on how deep you want the insulation to go. Clean the wall with water and a wire brush. Attach foam boards with either masonry nails or an adhesive. (Dow recommends Dow Mastic #11 for their Styrofoam TG brand material.) Petroleum-based adhesives cannot be used, and any adhesive must be compatible with the foam.

A number of materials are available to cover the foam once it's in place. Usually a stucco-like mixture of liquid latex modified cement is applied. This can be brushed or swept for a rough-surface if desired, or you can mix conventional premixed mortar mix with liquid latex. See your insulation dealer for advice. Any siding can be used over foam boards, but you have to use nails long enough to accommodate the extra depth.

The interior of foundation walls can also be insulated. In crawl spaces, you can attach batts to the band joist and let them fall onto the ground area some two feet from the wall. Use bricks or boards to hold them down. An ideal time to insulate foundation walls is when you're converting a basement into living space. One method is to build a stud wall just inside the masonry wall, using two-by-fours, and install batts or blankets between the studs. The vapor barrier, of course, should always face toward the "warm in winter" side.

One of the best and easiest materials I've found for this and other "problem" jobs is the R-Max Thermawall system. These "sandwich" panels come in several thicknesses (depending on R-value desired) with polycyanurate foam board faced with a full foil vapor barrier laminated to dry wall material. Metal bars are fastened to the wall using masonry nails. Adhesive is placed in

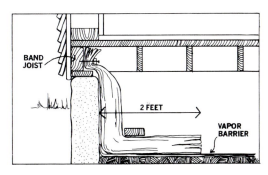

**Batt insulation** can be used to insulate foundation walls in crawl space by attaching to band joist, then allowing blanket to drop to ground area. Insulation should extend two feet from perimeter of foundation wall. Masonry or framing can be placed on blanket to hold in place at ground level. Be sure that a polyethylene vapor barrier is installed over the exposed earth.

**Exposed foundation walls** can account for ten percent of a home's heat loss, according to a Dow Chemical study. In photo above, Styrofoam board has been attached to foundation wall with masonry fasteners, mesh lathe is being installed. Foam boards must always be covered with fireproof material.

**Thermawall installation** begins with nailing metal bars to the wall and filling the channels with adhesive, as shown. Then the panels are pressed into position, and clamped with retaining clips. With panels in place, you have an R-19 finished wall that is ready to be taped and painted. Since there are no nail holes to fill, the finishing is easier than with conventional dry wall.

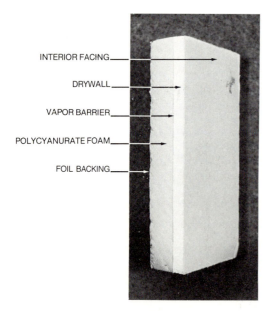

INTERIOR FACING

DRYWALL

VAPOR BARRIER

POLYCYANURATE FOAM

FOIL BACKING

**The R-Max Thermawall system** is an ideal way to insulate and finish a basement wall. As can be seen in the illustration, the sandwich consists of a polycyanurate board, with aluminum foil vapor barrier, laminated to dry wall.

**Sole plate** where foundation meets wall is prime source of energy leaks. Foam sealant can be used on inside or outside wall. When used outdoors, it should be shielded from direct sun.

channels on the bars, and then the panel is put in position and secured with special clips. In an amazingly short time you have a finished, insulated wall, leaving only the

seams to be taped as with conventional dry wall construction. And even that's easier because there are no nail holes to fill.

**One place to look** for sources of air filtration is anywhere a pipe or conduit enters your home's foundation walls. In these spots, you can use a foam sealant instead of caulking to provide both insulation and sealing. In some plumbing applications, the foam can also stop vibration and rattle.

## INSULATION-PLUS

After your insulating work is done, take another look at other potential sources of heat loss. Some things to consider:

- Storm windows. Are yours in good condition and tighty sealed?
- Switch plates or receptacles on exterior walls. Special "gaskets" that go behind cover plates will stop any air infiltration around them.
- Basement or attached garage doors. Insulating them can block air infiltration and stop heat loss.
- On the market you will find super quality storm doors, with foam-filled metal parts and thermal barriers between glass and frames, and insulated metal entrance doors that look so much like wood paneled doors that it's hard to tell the difference.

# Insulating wall coverings

*Energy savings with a decorator look*

## By V. Elaine Smay

### THIS ONE BOOSTS THE R-VALUE

It started when Floyd Baslow walked into an elegant hotel in London. "The walls were covered with fabric, and I noticed this gimp around the edges," he recalls. "I touched the gimp, and the tape that held it on fell off, revealing a row of nails beneath. There was a lot of confusion," he adds, "so I got out quick."

Thus inspired, Baslow set out to design a system for applying fabric to walls that wouldn't leave nail heads to be covered. The result is Fabri Trak, a PVC frame that grips the fabric. Under the fabric is a ½-inch polyester pad, which increases the wall's R-value by 1.41. It's also excellent sound-proofing. The noise-reduction coefficient is increased by an average of 118 percent when Fabri Trak is put over a concrete wall, according to tests, and by 374 percent with a gypsum-board wall.

Installers are trained by Baslow's company, Unique Concepts, Inc. (59 Willet St., Bloomfield, New Jersey 07003), and many

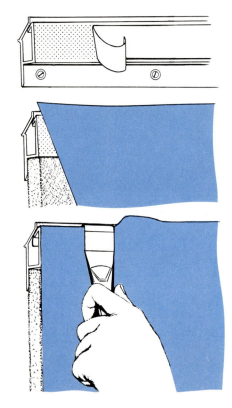

work through Sears, Montgomery Ward, J. C. Penney, and Ethan Allen stores. They put the frame around all edges of the walls and all openings, put up the polyester pad, and attach the fabric to an adhesive strip on the frame. Then, using a special tool, they insert the fabric into the frame.

Upholstered walls using the Fabri Trak system need no surface preparation, even if cracked. Nearly any fabric can be used.

## THIS ONE REFLECTS HEAT

A German-developed wall covering called Thermodecor blocks heat loss through walls by reflecting infrared radiation (radiant heat). The reflection element in the covering is a thin layer of pure aluminum. It's coated with decorative paint that reflects visible light (therefore you see it) but is translucent to infrared radiation. The result: Thermodecor looks like conventional wallpaper, with printed patterns and embossed textures, but reflects 65 to 75 percent of the radiant heat that strikes it (depending on pattern and texture).

Just how much energy you could save by papering a room with Thermodecor depends mainly on your heating system. The more radiant heat it emits, the greater the potential for saving. About 50 percent of the heat from radiators in a hydronic system is infrared radiation, according to Enertec Systems, Inc. (Box 127, Barrington, Illinois 60010), U.S. distributor of Thermodecor. But a smaller percentage of the heat from a forced-air system would be infrared radiation.

In tests conducted at the University of Munich, Germany, four walls and the ceiling of a small test room were papered with conventional wallpaper, which reflected 13 percent of the infrared radiation, then with Thermodecor, which looked identical but reflected 65 percent of the radiant heat. The room was heated to 21 degrees C (70° F) with three devices: a radiator, a forced-air heater, and a radiant floor heater. With a radiator as the heat source, 15 percent less energy was required to maintain room temperature with Thermodecor on the walls. With forced-air heat, however, Ther-

modecor reduced energy consumption by only five percent. The wall covering had the greatest effect with the radiant floor heater, which used 18 percent less energy when the room was papered with Thermodecor.

Reflective wall covering can also save much of the heat given off by objects in a room, which emit infrared radiation after they warm up, and it will reflect back your body heat. The reflective covering makes a room warm up more rapidly, says the company, so it could be a benefit in a room that you heat only for occasional use.

Test marketing of Thermodecor began in early autumn. It is applied like conventional wallpaper. Thermodecor-faced ceiling tiles and vertical louvered blinds are to be introduced later.

**Reflective wall covering** looks conventional to the eye (above left) but not to an infrared camera, which shows that it reflects the body's infrared radiation back into the room (far right). Ordinary wallpaper (center) shows little reflection.

# 7 WEATHER STRIPPING

**C**racks and crevices are always with us, and new ones are formed from time to time. It's important to keep them sealed; they not only allow precious warm, conditioned air to leak from the home, but are the source of unpleasant drafts as well. If you don't think a small crack is important, consider that a small opening that you would barely notice around a typical exterior door is equivalent to a 4-inch hole in the wall!

Thankfully, there's a wider range of good weatherstripping materials available today than ever before. This chapter gives you details about applying them to your home.

# The art of weatherproofing your house with weatherstripping

## Part 1: Doors

By V. Elaine Smay

Photos by A. J. Hand

If all the gaps around your exterior doors, all the cracks around your windows, all the crevices around kitchen and bathroom vents, room air conditioners, basement windows, and garage doors were lumped into one area, they would probably add up to a pretty big hole in your house. Spread around as they are, however, it's easy to forget just how much warm air escapes through these spaces in winter and how much cold air gets in. During the air-conditioning season, the reverse happens—with the same effect: higher-than-necessary utility bills.

The Canadian Department of Energy, Mines, and Resources reports that up to 25 percent of a home's heat loss may be due to excess infiltration around doors, windows, and other cracks. Drafts not only cost money, they also make you feel chilly. And that may cause you to turn up the thermostat and burn even more fuel.

Now the good news. Closing all these gaps is fairly simple. The Department of Housing and Urban Development figures that to seal a typical house would cost from $75 to $105. Yearly savings in heating costs, in all but the warmest regions of the country, should range between $40 and $100. If you have air conditioning, your summer savings will probably range from $67 to $167 a year, according to HUD. As a bonus, you'll probably shut out a lot of dust and noise, and stop any rain or snow leaks.

There are many weatherstripping products on the market. Just about any is better than none at all. Felt strips, closed- and open-cell foam tape, usually adhesive-backed, are inexpensive, available in various sizes, and easy and quick to install. But they are temporary solutions. If you want long-lasting weatherstripping, look for metal, vinyl, or combination products such as we show on these pages. (This is not a complete guide to all that's available, however.)

The products we show are intended to suggest the range of approaches to various problems. A trip to your hardware or home center will introduce you to the variations available locally.

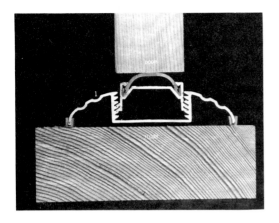

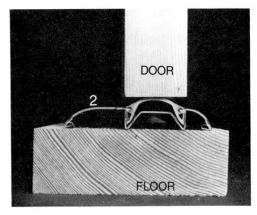

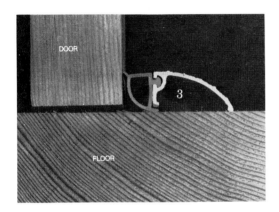

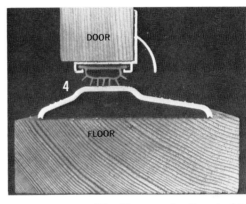

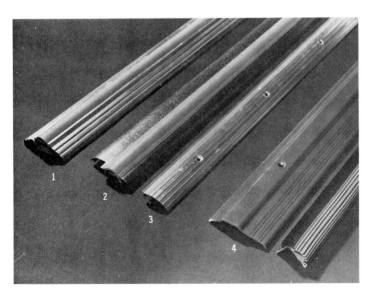

**Weatherproof thresholds** can seal the crack at the bottom of your doors. Photo at left shows four kinds; profiles above illustrate their use: 1) Aluminum threshold from Climaloc has vertically adjustable vinyl insert for snug fit. 2) Steel-and-vinyl unit (United Industries) is notched to fit door frame. 3) Bumper aluminum/vinyl threshold (Climaloc) makes contact with face rather than bottom of door; keeps out driven rain. 4) Replacement vinyl strips from Pemko puts vinyl strip on door bottom where it's not in foot traffic. Drip cap sheds rain.

223

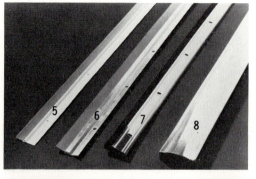

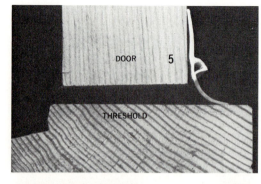

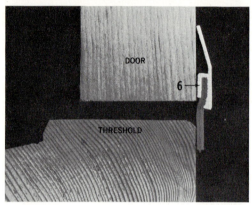

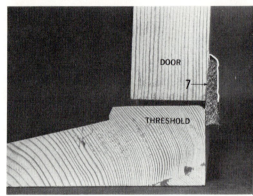

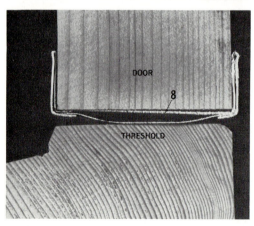

**Door-bottom seals** are easy to install: 5) Rigid vinyl strip has flexible vinyl seal that drags on threshold as shown. Deflect-O and Portaseal make these; Stanley makes a wood-and-vinyl unit. 6) Aluminum and reinforced-rubber door sweep (Climaloc) butts against threshold, as does 7) aluminum-and-felt unit from Macklanburg-Duncan. All have slotted screw holes for easy height and adjustment. 8) Brass door shoe (Kel-Eez) slides on door bottom. Door clearance must be 3/16 inches; shim can be used.

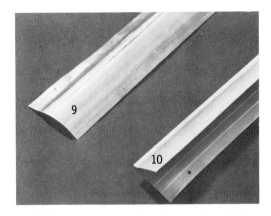

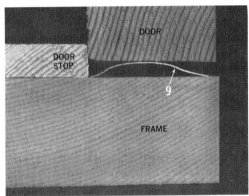

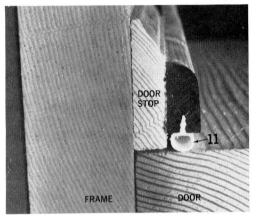

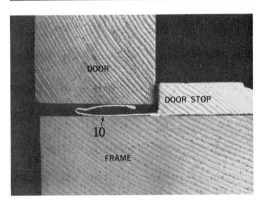

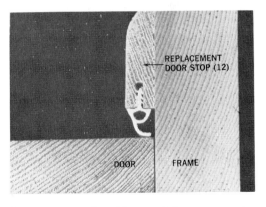

**Sides and top of door** can be permanently weatherstripped with spring metal: 9) U-shaped brass (Kel-Eez) has nailing flange on edge; springy curved surface snugs up against door perimeter. 10) V-shaped bronze (Macklanburg-Duncan) fills larger gaps. Jim Walter makes a V of aluminum.

**Door-stop seals with wood** frames can be finished to match door frame. Vinyl gaskets in these examples (Portaseal) are glued into channels of frames for durability. 11) Add-on seal can be quickly nailed to present door stop (or used around windows). 12) Replacement door stop requires that old one be removed.

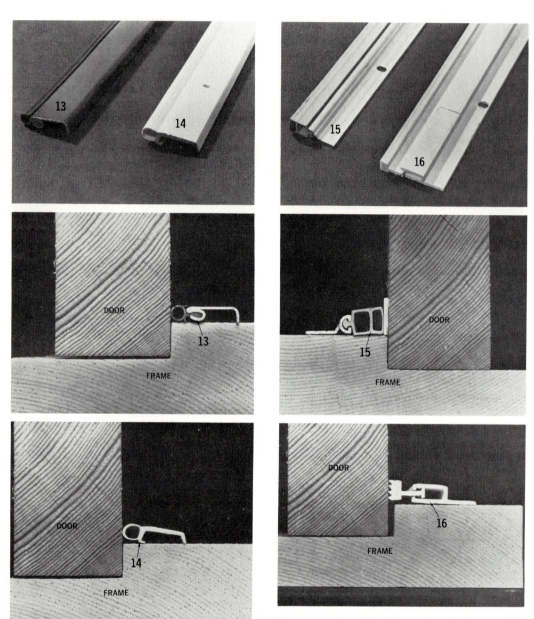

**Door-stop seals with metal and vinyl** frames are also available. Slotted nail or screw holes allow accurate adjustment along door perimeter. 13) Aluminum strip from Pemko comes in aluminum, dark bronze, or gold colors. 14) Rigid vinyl frame with flexible vinyl gasket installs with nails (Portaseal).

**Super seals:** 15) Magnetized door sets, similar to refrigerator-door gaskets, have aluminum and magnetized-vinyl strips for door stop that grip magnetized tape on door face (not needed on metal doors). 16) Spring-loaded vinyl strip adjusts to temperature, humidity, accommodates warpage. Both Climaloc.

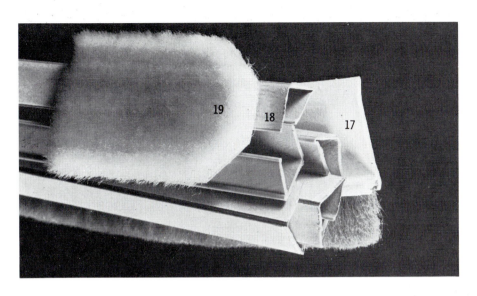

**Door-sealing kits** contain all you need to weatherproof a door. This from Schlegel includes 17) polypropylene pile door sweep, 18) Polyflex adhesive perimeter seal (in black, beige, white), and 19) polypropylene pile corner seals.

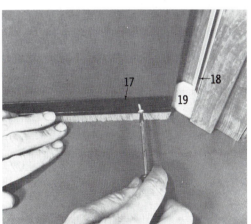

# The art of weatherproofing your house with weatherstripping

## Part 2: Windows, storm windows, garage doors

By V. Elaine Smay

"When I was a kid we stuffed newspapers around the windows and kicked the braided rug against the door at night," says Mike Meryn, president of Portaseal, a weatherstripping manufacturer in New Jersey. With today's energy costs, such a down-home approach to reducing air infiltration into your home just isn't good enough. The outside air that sneaks in—and the inside air that leaks out—can short-circuit your insulation and add significantly to your heating and cooling costs.

Check your windows for excessive air infiltration on a breezy day. If you feel air leaks, examine the glazing compound around the glass. Replace it if it has cracked badly or chunks have fallen out. Next, check the joint between house and window frame; caulk there if necessary. Then weatherize the perimeters of your windows with weatherstripping that doesn't interfere with their operation or

with the lock function. Window weatherstripping is sold both by the running foot and in kits for one or two standard-size windows.

Double-hung windows (the conventional type with one or two vertically sliding sash) require weatherstripping that can stand up to the friction their operation entails. Horizontally sliding windows present the same problem. Most weatherstripping made to install on the frames of double-hung or sliding windows can be applied either on the interior side or on the outside, but outside application may be best for appearance's sake.

Wood-frame storm windows should be weatherstripped around the perimeter where the frames butt against the window molding. Self-sticking foam tape will do the job, as will felt strips, but felt tends to deteriorate when exposed to the weather. When weatherstripping storm windows, be sure you don't close the weep holes at the bottom. These allow condensate to run off. Most aluminum storm windows come with replaceable pile weatherstrip (see photos).

FRAME

STRIP

SASH

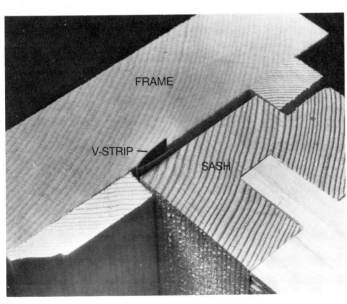

FRAME

V-STRIP

SASH

**V-shaped, self-adhesive plastic** installs easily, even on double-hung windows. V-shaped metal can be used if nailed in place. V-strips are placed in window channels as shown and on bottom face of upper sash to seal center crack. Another strip can be put on sill to seal bottom. Climaloc's Climaflex (above) is sold in three-foot strips, three to a set. Polypropylene strip from 3M (left), for wood or metal frames, comes in 17-foot rolls.

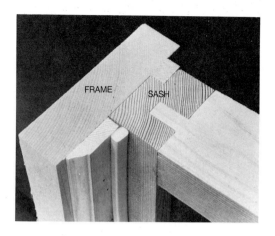

**Foam strip** in anti-friction, adhesive-backed vinyl sleeve (Climaloc) is applied to the inside frame for quick, inexpensive, nonabsorbent seal on wood or metal windows.

**Tubular vinyl** (from Elgar) is tacked or stapled to wood frame. Aluminum and wool-pile strip made by Lustre Line (not shown) can be used similarly on metal windows.

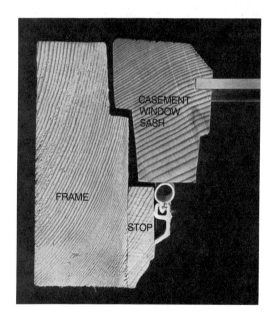

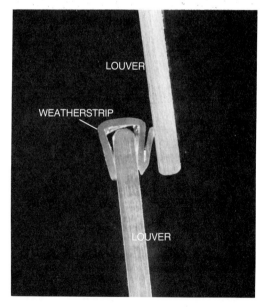

**Hinged windows can be weatherstripped** in many ways. Here, Lustre Line's soft vinyl seal in aluminum frame is attached to window stop. Other types are put between stop and window frame and are invisible with window closed.

**Many narrow louvers on jalousie windows** mean many feet of cracks for infiltration. Stop it with clear vinyl weatherstripping (this from Pemko) that you cut to length from a 25-foot roll and snap over the edge of each louver.

230

**Large gap around garage door** requires special weatherstrip, usually sold in kits. Weatherstrip shown, from Portaseal, has rigid vinyl nailing strip with wide flexible vinyl seal (below left). Ends are notched for weathertight overlap at top corners (center). Below is rubber seal from Manco Tape. Other types are made.

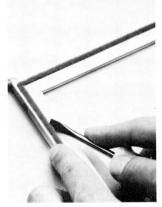

**Replacement pile weatherstrip** of long-lasting polypropylene for aluminum storm windows (and doors) is available from Stanley and Schlegel. Check present weatherstripping to see if it has shrunk, matted, or pulled back from corners. If so, pull it out and press replacement strip into same slot with screwdriver or knife. Cut to length with knife or scissors. New pile should block entry of rain and dirt, stop window rattling, and keep out cold-air drafts.

# Storm-window kits that outperform regular storms

*The one-piece design of these cheap, easy-to-install kits stops drafts cold*

*By A. J. Hand*

When do-it-yourself storm-window kits came out a few years ago, their main attraction was easy assembly and installation. The plastic glazing material and vinyl-frame moldings were easy to cut with ordinary tools. Inside installation eliminated the hassles and dangers of working outdoors on a ladder.

But now the kit makers have discovered an even more important asset: excellent thermal performance. According to recent tests, do-it-yourself storm windows can outperform conventional triple-track combination storm/screens by a significant margin.

The reason? They do a far better job of blocking air infiltration. Triple-track storms are much more convenient to use than the kits. They let you convert from storm window to screen and back simply by sliding the glazed sash up and the screened frame down, and vice versa. But this convenience comes at a price. It's difficult to provide a truly tight air seal around these moving parts. Cold drafts can slip in around the edges of the storm sash, especially at the joint where the upper and lower sashes meet.

Storm-window kits cover your entire window with a single sheet. Since there are no moving parts, sealing against infiltration can be nearly perfect. In tests conducted by Building Research Associates for Plaskolite, the In-Sider kit storm window cut infiltration losses by more than 98 percent in a Boston test house, by 89 percent in Detroit, and by more then 92 percent in Philadelphia. In these same test houses, triple-track storms reduced losses by only 34 percent in Boston, 6.2 percent in Detroit, and 18.3 percent in Philadelphia.

Keep in mind that these figures are for infiltration only, not for heat lost through the glazing via conduction and radiation. In preventing heat loss through conduction and radiation, all storm windows perform just about the same, although ordinary aluminum storms are about 20 percent less effective than wooden storms or aluminum storms with a thermal break. Still, those figures for infiltration are impressive.

How much fuel and money can the kits save? That's hard to predict accurately, because infiltration is a function of so many variables. Inside and outside temperature, wind velocity and direction, and the overall tightness of your windows all affect infiltration economics. If your windows are a tight, modern design, they will permit very little infiltration, and you might be happier with the convenience of combination storm windows. But if your windows are loose, counterweight clunkers that rattle in the wind, infiltration becomes a big consideration, and the kits have a real advantage.

## WHAT'S AVAILABLE?

Storm-window kits come in a variety of types, each with its own set of advantages and disadvantages. Here are some of the major brands:

- The In-Sider features a self-adhesive vinyl frame that you cut to size, fit with acrylic glazing, and press in place on your window casing. It's for interior use only. The adhesive bond between frame and casing provides a good seal. Since the frame can't be removed without destroying the adhesive, the In-Sider frame is designed to snap open and release the glazing for ventilation during the summer.
- The Barrier is another self-adhesive vinyl system. The frame snaps apart to release the glazing, usually 0.080-inch acrylic.
- Therma Frame also uses a vinyl frame that you cut to size, but in this case the frame installs with turn buttons that screw into the window trim, either inside or out. It can take glass or acrylic, but ordinary single-strength glass is a safety hazard when used in sheets as large as an entire window.

**Kit storm windows** can be installed indoors without the need to climb ladders—a safety advantage on upper-floor windows. Tests show that storm windows can reduce infiltration by over 98 percent in some climates.

**Therma Frame,** Mr. Window, and Thermatrol windows all fasten in place with turn buttons. In-Sider uses pressure-sensitive adhesive on the back of the frame; the frame snaps open to allow easy removal of the glazing.

Therma Frame mounts directly against the face of your window trim—hard vinyl against wood. You can probably improve its seal by applying a narrow strip of foam weatherstripping around the rear face of the frame before mounting it.

**Note:** All three kits involve a system of three frame parts. One is the basic frame extrusion. Another lets you join two storm panels together to cover large windows or glass doors. The third is a special strip to provide a seal along the windowsill (see photo).

• Thermatrol Stock-Line windows use the same sort of three-part system as the windows mentioned above, but operate more conveniently. The windows are made in stock sizes, precut and ready to assemble and install. The 0.065-inch acrylic glazing is already gasketed into the acrylic frame parts. The rear surface of the frame has a foam backing for a tight seal when the window is pressed into place with turn buttons.

The unique feature of the Thermatrol window is its availability in either a one-piece version that covers the whole window with a single sheet or a two-piece design. The two-piece design has upper and lower panels. Whenever you want ventilation, you can remove the lower panel and clip it over the top one for storage. That makes Thermatrols almost as convenient as combination storm windows, yet still provides a good seal against drafts.

• Mr. Window uses precut frames of aluminum rather than plastic. You supply your own acrylic glazing, assemble it in the frame, and secure it inside or out with turn buttons. The glazing is secured with a vinyl spline, which will also secure screening, letting you convert the frames for summer use.

## PRICES

You'd expect the kits to sell for less than aluminum combination storm windows, and most of them do. Although prices vary considerably (since many home centers sell at a discount), here are some rough figures for windows in the 30-by-55-inch range.

Therma Frame with 0.080-inch acrylic glazing is about $15. A one-piece Thermatrol is about $13, and the two-piece model about $18. Mr. Window with 0.080-inch acrylic would cost around $23, the same price I have paid for aluminum combination storm windows at a home center down the road from me.

## INSTALLATION TIPS

All the kits are easy to assemble and install, but a few tips will help make sure you get the most from your installation.

If you're using adhesive-mounting windows, be sure your window casing is clean and free of dust, wax, or grease. These materials interfere with good adhesion, and once the adhesive loses its effectiveness, much of the weather seal is lost.

If you're using turn-button-installed windows, be sure to use enough buttons. Both the vinyl frames and acrylic glazing are a bit floppy. Space the turn buttons every six inches or so to insure a good seal.

Inside or outside installation? Performance should be about the same wherever you put the storm window, as long as you caulk around the perimeter of your exterior window trim. Exterior mounting gets the nod in terms of aesthetics, but interior mounting makes for easier installation, cleaning, and removal.

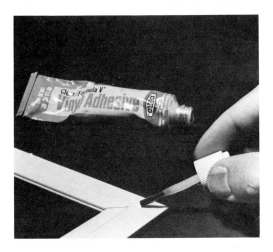

**Therma Frame** (top left) can be mitered or butt-joined at the corners. Parts included in the kit are frame strip, joiner strip for joining two storm windows together, and sill strip to secure the bottom edge of the storm window along the windowsill. Vinyl Therma Frame can be cut with back saw in a miter box (top right). It will accept glass or acrylic glazing in 0.080- or 0.100-in. thickness; acrylic is tougher, safer, and easier to cut to size. A dab of vinyl cement at the corners (lower left) will lock the frame together and keep it from accidentally slipping off the glazing during installation or removal of the window. Storm windows should be left on year-round if your home is air-conditioned. Therma Frame can be installed outside if you like and painted with latex trim paint to match the color of your home (lower right).

## COLD FRAME

Storm windows are not the only application for these do-it-yourself kits. I used Therma Frame to glaze my garden cold frame. Installation is easy, and the acrylic glazing I used is tougher than glass. If it ever does break, repair is simple—just slip new glazing into the vinyl frame and fasten it back in place with the turn buttons. Though I used Therma Frame, any of the other clip-on systems would work just as well. They all make any glazing installation a snap.

# SOLAR ENERGY FOR HEATING THE HOME

**U**sing the energy of the sun has great appeal—it's available everywhere and it's free. But the equipment to use it is often expensive to buy and maintain, and as you'll see here, the solar homes that have proven most effective are those that are rather simple in design.

Whatever you do, don't overlook the principles of solar energy, for you can take advantage of them in any program of improving your home's energy consumption. It may be no more than installing shades or even a window in the proper location, or orienting your home to the south and installing the major glass areas there; it may mean a major solar project such as a solar attic or complete home. In any case, making the best use of the sun offers something for everyone.

We are grateful to *Homeowners How-To* for the use of material appearing in "If you understand the basics, solar energy can work for you"; "Solar collectors"; "How solar heat is stored"; "The distribution system"; and "Now: A do-it-yourself solar heating system."

# Solar heating:
# How to pick the right system
# for your home

*Here's an inside look at how a pro might design and install a solar heating system*

## By Richard Stepler

The slide that flashed on the screen showed a contemporary residence in the mountains above Los Angeles. A large, dramatically angled roof readily identified it as a solar home. But there was a catch: The roof was covered with shingles, and the solar collectors were mounted in racks on the ground at right angles to the house.

"That," said solar engineer Jim Senn, "is how *not* to build a solar home. The roof pitch and area are fine, but it's facing in entirely the wrong direction."

I was attending a two-day seminar in Los Angeles to learn how professionals plan and install solar heating systems. When Jim Senn isn't giving courses to other dealer/installers, he's in the field with his own solar company, installing Energy Systems, Inc. solar-heating equip-

ment in the Los Angeles area. He was called in too late to save the owner of the "wrong-way solar contemporary" some unnecessary expense.

That's lesson number one in buying a solar heating system: Get advice from a pro. And, just as important, first learn the basics yourself.

## THE BASICS

Solar heating is deceptively simple. The sun heats up a blackened absorber plate that is usually covered with a sheet or two of glass or plastic and insulated on the back and sides. A liquid, circulated in tubes attached to the absorber, or allowed to trickle down its surface, or air ducted to it carries the heat where it's needed: to heat water or living space right away; or to heat a tank of water or bin of rocks for use at night or when the sun's not shining.

The difficulties arise in choosing a system that's properly sized to your requirements; selecting components that perform well in your region's temperature extremes; and getting all the elements to work together. In addition, if you're retrofitting, you must select a solar heating

system that is compatible with your existing water- or space-heating equipment.

The three most common applications of solar energy for the home are domestic water heating, space heating, and swimming-pool heating. Here are general guidelines for installation of these three systems.

## SOLAR WATER HEATING

The drawings on the following pages illustrate four types of solar systems that will heat water for domestic use. In addition, air-type collectors can heat water via an air-to-water heat exchanger. (Solaron Corporation makes such a system; see drawing under "Solar Space Heating.")

How large a system do you need? The average U.S. family uses from 20 to 35 gallons of water per person per day, depending on habits, or a rough average of 30 gallons per person. "ESI's double-glazed collector is sized to produce 30 gallons of hot water per day," says Senn. So you will need one collector for each person in your family.

Recommended storage capacity is also 30 gallons per person or per collector. In two-tank systems the backup water heater's capacity is included in the total. A system that's properly sized to your family's needs will supply at least 70 percent of the energy used to heat water.

"Usually water in a solar heating system will be hotter than in a conventional one," says Senn. "It's possible to get temperatures from 165 to 185 degrees—dangerously high for skin contact." ESI's systems use a tempering valve to reduce water temperature to a safe level (see drawings).

# Solar water-heating systems

**Small number of collectors** (two to four is typical) and year-round use make this solar application cost-competitive with conventional fuels. Simplest system, with no energy input, is thermosiphon; it requires that storage tank be located above collectors. Where this isn't possible, a pump controlled by a differential thermostat circulates water.

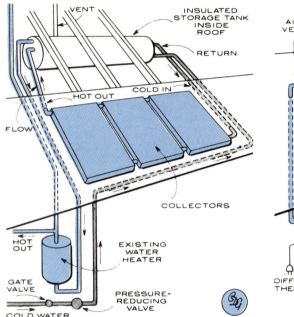

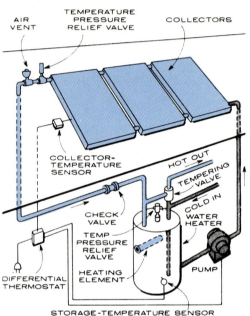

**In thermosiphon,** or natural-circulation system, denser cold water falls, absorbs heat in collectors, and rises to storage tank. To prevent heat loss by reverse flow at night, tank must be at least one foot above collectors. In areas where freezing temperatures occur, a nonfreezing solution can be circulated using the same principle with heat exchanger in tank. (System shown: Sun Power, New Zealand.)

**Single-tank** pumped system has backup electric heating element near top of tank. It's intended for new installations or when existing water heater must be replaced. System shown (Energy Systems, Inc.) is direct-heating type; potable water from bottom of tank is heated in collectors and returned to top of tank. In case of freezing temperatures, water automatically drains out of collectors.

Other tips from Senn:

• Insulate all hot-water piping runs, including those to and from the collector(s).

• Add insulation to the storage tank(s). (A major reason that many of the systems in New England Electric's 100-home test performed poorly was that they were underinsulated, claims Senn.)

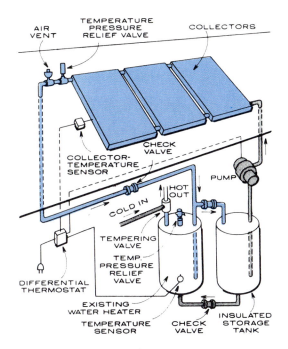

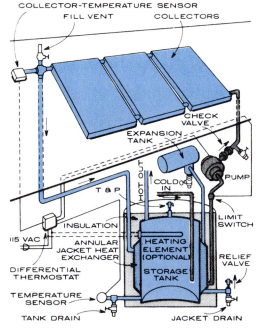

**For retrofits,** existing electric or gas-fired water heater can be part of system. Shown is full-circulation system; a loop connects tanks so sun can heat all the water in both. Solar-heated water can be hot: 185° F. Tempering valve mixes cold and hot water to keep temperature at safe level (typically 102° to 140° F). ESI direct-heating type here automatically drains in case of freezing.

**Indirect** or closed-loop system heats water in storage tank via heat exchanger that's either inside tank or wrapped around it (shown). Nonfreezing heat-transfer fluid needn't be drained from collectors. Local codes often specify double-walled heat exchangers when a toxic antifreeze is used in a potable water system. Indirect systems can use one tank (shown) or two tanks.

---

- Pumps (if any) must pass code for use in a potable water system. They must have nothing but bronze or stainless-steel parts in contact with water. They should be quiet, have low horsepower, and low speed. A pump for a water-heating system might draw only 30 watts and be rated at 1/220 horse-power. (Compare that to the 4,500-watt calrod you might otherwise be heating the water with!)

- Air vents are required at all high points in the system.

- The storage tank should have a pressure/temperature relief valve.

- Check valves, which permit flow in one direction only, prevent hot water in the storage tank from thermosiphoning to cool collectors at night and losing heat.

  Senn recommends that the temperature differential between collectors and storage be set at 20 degrees (DT on) before the system turns on, and that DT off, or the temperature difference between storage and collectors that shuts the system down, be set at three degrees.

If the system circulates potable water (not a nonfreezing fluid) through the collectors, as ESI's do, it must have a reliable freeze-protection capability. ESI provides a special differential thermostat (Rho Sigma's RS500 PH2L) and two solenoid valves, one normally closed (takes three watts to open) and the other normally open (takes three watts to close). As ambient temperature approaches freezing (or if power fails), the valves close and open, respectively, causing water to drain from the collectors at any point in the system where freezing could occur. This amounts to only a few quarts at the most.

If the system heats water indirectly with a nonfreezing heat-transfer liquid (such as an ethylene or propylene glycol mixture or a silicon-based fluid), it needn't have this feature. This type of system requires a heat exchanger, either wrapped around the storage tank (shown in drawing) or inside it. Local codes often specify double-walled heat exchangers when a toxic fluid is used in a potable water system. Some manufacturers recommend using nontoxic propylene glycol with a single-walled heat exchanger. You should check your local code first before you make a decision. Also, a heat exchanger reduces heating efficiency, and a system of this type requires a larger collector area.

## SOLAR SPACE HEATING

The drawings illustrate two types of solar space-heating systems: one in which air transfers heat to a rock storage bin; and one in which a liquid transfers heat to storage, which is most often a tank of water. A hybrid system uses an uninsulated tank of water buried in a bin or rocks for storage. Collectors heat the water in the tank, which in turn heats the rocks. Room air is ducted to the rock bin for heating.

Air systems are not subject to freezing, leaks, or corrosion. The rock storage bin acts as its own heat exchanger, and temperature stratification is good. Air blown in at the top at 140 degrees loses its heat as it passes through the rocks and is returned to the collectors from the bottom at 70 degrees. This assures efficient operation of the collectors. Air is a poor heat-transfer medium compared with water, however. Also, air ducts are larger than pipes, and blower horsepower requirements are higher than those for pumps in liquid systems. You must consider all these factors before you make a decision.

It's more difficult to size a solar space-heating system to a home than it is to select a solar domestic water-heating system. The solar heating system is sized to replace a certain proportion of heat that is lost from the home. There are well established formulas for determining heat-loss characteristics, but they are complicated and are probably best left to a solar engineer.

"Every house has a personality," says Senn. Its site, its construction, whether it's built to take advantage of natural solar gain all are factors in choosing solar equipment. If you're retrofitting, you may have space for only a limited number of solar collectors, or for a small storage tank.

As rules of thumb, Senn offers these guidelines:

- Don't try to provide 100 percent of the heating requirements by the sun. And don't design to worst-case conditions. A system that supplies all your heat in January, for example, will be dumping excess heat the rest of the year. It would be too large and expensive.

- Don't skimp on insulation. A solar-heated home should be well insulated for its climatic zone and have minimal infiltration losses.

## Solar space heating

**More solar collectors,** depending on your home's heating requirements, and larger storage (three to five days' worth is usually recommended) make this solar application a major investment. Most also provide domestic water heating to maximize return on investment. Two types— air and water—are shown below.

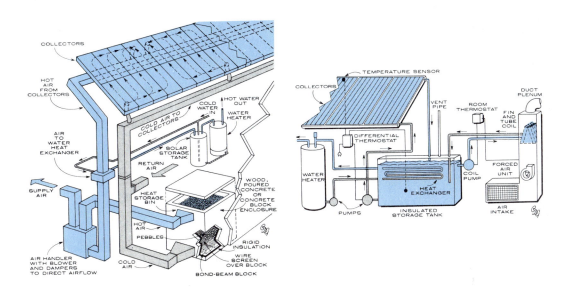

**Solaron's air-type system** (above, left) minimizes external duct connections: There's one inlet and one outlet for every eight collectors (156 sq. ft.). Air handler with blower and motor-driven dampers directs air flow throughout system. Heat storage is bin of ¾- to 1½-in. dia. pebbles (rule is ½ cu. ft. of rock per sq. ft. collector) with vertical air flow. When heat is stored, air flows from top to bottom so that coolest air is returned to collectors. To heat from storage, flow is from bottom to top so that hottest air goes to living space. In summer, rock bin can be cooled at night to provide daytime space cooling. Air-to-water exchanger supplies hot water year round. Energy Systems, Inc. liquid-type system (above, right) illustrates one of many possible variations. Storage water is circulated through collectors, which automatically drain when temperature drops to within a few degrees of freezing. An alternate system circulates a nonfreezing fluid through the collectors and a heat exchanger immersed in the storage tank. Space heating can be via water-to-air heat exchanger in forced-air system (shown), or heated water can be circulated in hydronic space-heating system. Tap water is heated via heat exchanger near top of storage tank.

• Leave room for expansion or modification of the system—especially in the collector array. If a system doesn't deliver the Btu's needed, you can add collectors.

## SOLAR POOL HEATING

Using the sun to heat an outdoor swimming pool is a simpler matter (see drawing). As a rule, collector area should be equal to at least 50 percent of the pool's surface area; but a pool in a shady or windy site might warrant up to 75 percent collector area to surface area.

Pool collectors are simply unglazed, uninsulated absorber plates, because the desired end-use temperature (most people prefer 80- to 85-degree water for swimming) is often at or below ambient air temperatures. In fact, a properly performing pool collector will feel cool to the touch.

Pool water should be circulated at a high rate. Senn recommends six gallons per minute through each collector. Typical pool circulator pumps, necessary for filtration, are able to provide the needed flow rate.

After I attended Jim Senn's seminar, I visited several Energy Systems, Inc. solar installations in the San Diego area. They ranged from retrofits on existing homes for domestic hot water and swimming pool heating to new homes specifically built for solar heating. Collectors were mounted on the roofs, on racks on the ground, even a patio sunshade. There were small and medium-size residential pools, and a large pool that served a whole subdivision. I was impressed with the variety of installations I saw; it seems that the means of using solar energy vary as widely as individual needs and the constraints imposed by existing buildings, topography, and local codes.

## Solar swimming-pool heating

**High flow rate through unglazed,** uninsulated collectors is recommended for best performance in heating outdoor pools. Solenoid valve, spliced into pool's existing filtration plumbing, is controlled by differential thermostat. This device compares two temperatures: collector and water. When solar heat can be added to pool, valve closes and water is diverted to collector array.

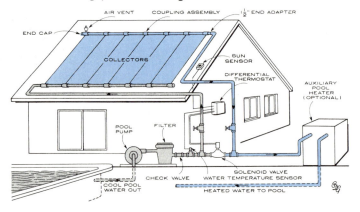

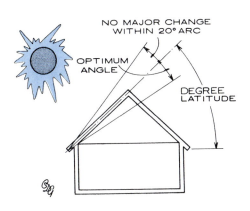

Collector tilt

Collector orientation

**Optimum tilt:** your latitude plus 15 to 20 degrees for maximum winter heating; latitude minus 15 to 20 degrees in summer. For year-round output, latitude ± 10 degrees is acceptable.

**Performance is not greatly affected** if collectors face as much as 20 degrees east or west of due south. In areas subject to morning fog or haze a southwest orientation may even be preferable.

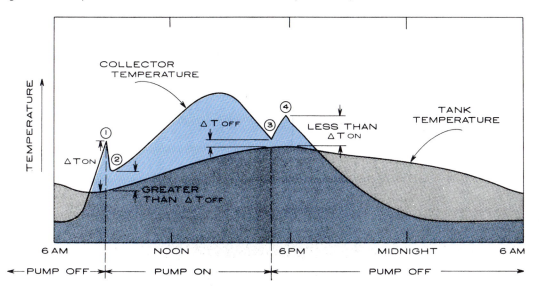

How a differential thermostat controls solar heat collection

**Sunny day scenario:** At sunup, pump (or blower) is off. Collector temperature rises above storage temperature until temperature difference (DT) is sufficient to start heat collection (DT on at point 1). Cool heat-transfer fluid causes collector temperature to drop to point 2, but since this is higher than DT off, system continues to operate and store heat. DT off is finally reached at 3, and system shuts down. Now heat is no longer being carried away, and collector temperature rises to point 4. Since this DT is less than DT on, system stays off. Without this provision, system would cycle on and off repeatedly.

245

# Solar attic house—
# Simple low-cost system
# for winter heating,
# summer cooling

*Unique two-chamber attic heats
and cools a home designed by U.S.
Dept. of Agriculture researchers*

## By Richard Stepler

Nestled in a Georgia woods is a home
you'd probably drive past without notic-
ing. In fact, there's nothing much out of
the ordinary about its appearance from
the street—it's a conventional one-story
ranch with a slightly unusual roof pitch.

If you were to walk around to the backyard, however, you'd quickly spot one of
the features that makes this home unique:
a large glassed-in roof.

The house is Solar Attic Prototype No.
2, located near Athens, Georgia. It is but
one example of a hybrid active/passive
solar-home design that offers several
exciting advantages:

- It's simple to build, using existing materials that are familiar to home builders.

- It needs no expensive solar collectors
attached to the roof; the home's attic
collects the sun's heat, providing up to
75 percent of its space-heating requirements.

- Its solar heating system can work in reverse to help cool the house in summer.

- Solar water heating is relatively inexpensive to add to it.

- Its solar system is low in cost, amount-
ing, say the designers, to only about 15
percent of the total cost of the home.

The solar attic is the brainstorm of a
U.S. Department of Agriculture (USDA)
design team—research leader Ted Bond,
architect Harold Zornig, and engineer
Luther Godbey. They're with the USDA's
Rural Housing Research Unit, in Clemson,
South Carolina.

The first prototype solar attic house
was built in Greenville, South Carolina by
Heliothermics, Inc. Prototype 2, shown on
these pages, incorporates many improvements in the design. But these homes are
only part of the story:

National Homes, the prefab-home
builder, has erected its own solar attic

**See-through roof** of Solar Attic Prototype is double-glazed with tempered glass and UV-resistant Llumar.

**The system components:** two-chamber attic, blower, heat pump, ducts, and rock storage (see text for how it works).

house in Fayetteville, Indiana. The company is monitoring the home for a year before deciding whether to offer it in its prefabricated-home series.

The Appalachian Regional Commission is sponsoring several solar attic houses throughout the Appalachians. Seven homes are under construction in this project.

Plans to build the solar attic house are available from USDA Extension Service offices throughout the country. Three variations in the design allow for differing climates.

In all, the solar attic house has six operating modes: three for heating and three for cooling. Control of the system is done by a differential thermostat, a device that compares various temperatures (collector, house, storage) and activates system components accordingly. Here's how it works:

## HEATING MODES

The solar attic is divided into two chambers (see illustration, page 247): The upper chamber serves as the heat collector. It's glazed to admit solar energy and sealed from the lower chamber, which acts as a return-air plenum. The heat pump's evaporator coil, which supplies heat when the solar attic can't, is located in the lower chamber. (The heat pump's condenser coil, of course, is located outside the house.

A ⅓-horsepower, four-speed air-distribution blower in the upper attic chamber is capable of moving 1,500 to 1,800 cubic feet of air per minute. When the sun has warmed air in the upper chamber, the differential thermostat starts the blower. It draws air from the lower chamber through a series of four- by 21-inch openings along the base of the upper chamber's glazed surface (one is shown in the cross section on page 252). A gravity-

operated polyester flap attached to the top of each opening prevents reverse circulation of cold air from the upper to the lower chamber at night.

Heated air is ducted directly to the house or to the rock bin in the crawl space under the house. Air-flow arrows in the illustration on page 252 show both modes. At night or during cloudy weather when no solar heat is available, stored heat is ducted from the rock bin to the home's living space by the heat-pump blower, but in the reverse direction from which it's stored. Here's why:

Heated air enters the rock bin from the upper attic chamber in the heat-storage mode at, say, 120 degrees. The rock it reaches first rises to this temperature. As the air travels horizontally through the rock bin, it gradually raises the temperature of the rocks in a right-to-left direction, until it exits via ducts for its return trip to the attic at, say, 70 degrees.

When you want to remove heat from storage you reverse the air flow so it moves from left to right. Then air enters storage at, say, 68 degrees, and exits at 120 degrees, since it reaches the hottest rock last. If storage is exhausted (tests indicate that storage is sufficient for three sunless days), the heat pump supplies heat.

"It's really a hybrid system rather than strictly active," explains architect Zornig. "Passive features are the large sealed and heated attic and the crawl space, which is also sealed and heated. The only heat loss is through the walls."

Early in the heating season the house overheated because the rock storage became overcharged and warm air leaked through floor registers. Gravity-type dampers were installed at each register to overcome this problem.

Maintaining a relatively airtight attic also proved bothersome. As wood dried, cracks developed, and air leaked through every crevice. Air also leaked through imperfect seals in ridge and soffit vents.

It's important that the homeowner pay close attention to caulking and sealing—especially during the first couple of heating seasons—if optimum system efficiency is to be maintained.

Despite these minor hitches, Zornig says that the system performed well during the first winter. "Attic temperature never became too hot, peaking at about 110 degrees during clear winter days," he reports. Collection efficiency peaked at about 68 percent on one relatively warm, sunny winter day.

## COOLING MODES

An air-intake vent on the north side of the upper attic chamber (visible in the front view of the house in the photo on page 250) is opened in summer to draw in cool, nighttime air. It's ducted directly to the air distribution blower (now isolated from the upper attic chamber), which either circulates cool air directly through the house or cools the rock storage.

During the day, cool air from storage is circulated through the house by the heat-pump blower. Air flow through storage is reversed for the same reason as in heating, except that in cooling, of course, you want air to encounter the coldest rock last.

Additionally, soffit and ridge vents are opened in summer to ventilate the attic and thereby reduce the home's cooling load. These function both by forced circulation—the blowers force return air from the house into the lower and upper chambers, respectively, and out through the ridge vent; and by natural thermosyphon—as air heats up in the upper chamber it rises and exits via the ridge vent. This draws in cooler air through soffit vents in the lower chamber. "The lower chamber," says Zornig, "stays 30 degrees cooler in summer than most conventional attics."

## SOLAR HOT WATER

Domestic hot water is preheated throughout the year in unglazed and uninsulated absorber plates placed on the floor under the glazing in the attic's upper chamber (see photo, page 250). Aluminized Mylar reflects additional insolation onto the plates. Maximum water temperature achieved in the test house: 140 degrees. If the attic drops to 35 degrees, the absorber plates automatically drain.

The 1,288-square-foot house is well insulated: Floor, walls, and ceiling have six-inch-thick fiberglass batts. There's an airlock (double-door) entry, and all windows are double glazed.

Prototype 2's collector consists of 381 square feet of double glazing angled at 50 degrees (latitude plus 15 degrees). Outer glazing is ⅛-inch-thick, low-iron-content, tempered glass. Inner glazing is a specially treated ultraviolet resistant polyester called Llumar made by Martin Processing Company of Martinsburg, West Virginia. It has a 10-year estimated life.

The tempered glass is set between the rafters on 2 x 2 spacers. Llumar is stapled to the inside edge of the spacers, leaving a one-inch air space. An alternate glazing, used in the first prototype, is translucent fiberglass-reinforced-plastic panels. The designers used glass and polyester in the second prototype because tests indicated that from 10 to 20 percent more solar energy would be transmitted through this combination. "Also," says Zornig, "tempered glass is now available at about the same cost as fiberglass-reinforced plastic."

Fifty tons of crushed rock (No. 4 railroad ballast), or one cubic foot of rock for each square foot of collector, provides heat (and cool) storage. Normal rule for rock storage in an air-type solar heating system is one half cubic foot of rock per square foot of collector. "The rock storage is sized for cooling," explains Zornig.

**Prototype 2's hot water** is preheated in uninsulated absorber plates placed behind glazed roof. Mylar reflects additional insolation onto plates. Note provision for manual clothes drying. From the front (right), only the unusually pitched roof gives the house away.

"Nocturnal cooling requires a larger volume for adequate storage." The additional storage is also useful in the heating mode, however, providing a longer carry-over of solar heat during cloudy weather.

"What we tried to do," says Zornig, "was to show how the system could be most economically built." Many variations are possible; one, designed by Aztec Solar Homes—the Athens, Georgia, based builder of Prototype 2—moves the collector to one side of the house.

"You're not stuck with that Pizza Hut type of roof," adds Zornig. "You could have almost any roof shape—as long as you provide the double-attic feature."

## $36,000 SOLAR HOME

Zornig estimates that the cost to build the test house *excluding* the solar components was from $28,000 to $30,000. "Solar-com-ponent costs," says Zornig, "break down like this: $1,100 for the air-distribution system; $1,000 for thermal storage; $564 for the solar water-heating system; and $3,000 to build the upper attic, including glazing." So total cost for the house was in the $33,000 to $36,000 range.

"Economic analyses we've run," says Zornig, "show about a 10-year payback for space and water heating alone." Any contribution by nocturnal cooling would shorten payback.

Solar economics, of course, will vary for each individual, depending on solar fraction (percent of heat supplied by solar), costs of conventional energy, mortgage interest rate, and income-tax bracket. Newly enacted federal legislation is another consideration. Also, many states and localities offer tax incentives and real-estate tax exemptions that could figure heavily in a solar cost analysis.

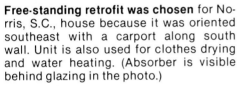

**Free-standing retrofit was chosen** for Norris, S.C., house because it was oriented southeast with a carport along south wall. Unit is also used for clothes drying and water heating. (Absorber is visible behind glazing in the photo.)

**House in Pendleton, S.C.,** had wall with two small bedroom windows facing 28° west of south, so attached retrofit was possible. Collector is angled at 50°, or latitude plus 15°. Solar water heater is included here, too.

Another project undertaken by the USDA design team involved retrofitting the solar attic system to existing homes. Two houses were chosen, each with 1,043 square feet of living space. One, near Norris, S.C., got a free-standing solar unit (photo at left). The second, near Pendleton, S.C., was equipped with an attached system (right).

First—essential for *any* solar retrofit—insulation was added to the ceiling and crawl space of each house, doubling thermal resistance. Storm windows were also installed. These improvements reduced the homes' heat demands by 23 percent.

The solar retrofits are constructed as shown in the drawings on page 252. The collector is angled at 50 degrees (latitude plus 15 degrees). There is one cubic foot of rock for each square foot of collector.

Two 1/6-hp blowers move the air; one, located in the collector, moves air from collector to storage. A differential thermostat starts in the blower when collector air is 20 degrees hotter than storage. The distribution blower, in the crawl space next to the rock bin, moves air to the house. This blower is controlled by a thermostat inside the home.

Manually operated vents in the collectors allow for summer ventilation.

Both of the prototypes became too hot during summer testing (one peaked at 120 degrees with a 2-degree ambient); the USDA researchers feel that larger vents should help solve the problem.

Solar water-heating systems are also included in the retrofits. The attached unit heats water by thermosiphon; hot water rises through a copper solar heat absorber into a storage tank that's placed above the absorber. Cold water from the bottom of the tank enters at the bottom of the collector to complete the cycle. Sixty-seven percent of the hot water used by the family was provided by this system during the test.

The free-standing retrofit uses a 1/200-hp pump to move water from absorber to storage. This system provided 73 percent of the hot water used during the test period.

USDA's study shows that solar water heating is cost effective for these retrofits. But solar space heating isn't. Preliminary data show that each system saved the homeowner only about $90 annually, against the cost of electric resistance heating at three cents per kWh. This saving does not justify the cost of the solar retrofits. The experiment is continuing, however, with the aim of improving performance and reducing costs.

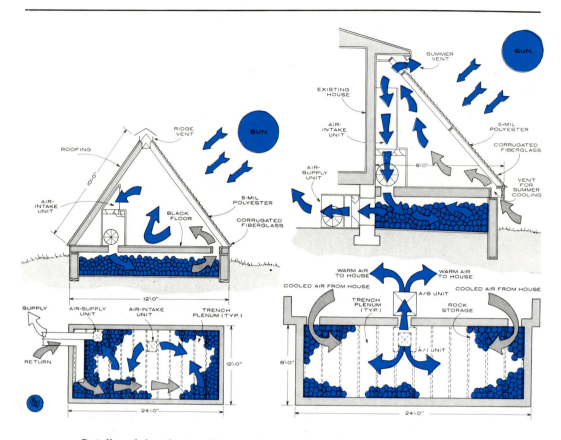

**Details of heating/cooling modes** of solar attic house are seen in illustrations above. See pages 248 and 249 for explanation.

# If you understand the basics, solar energy can work for you

*Start this short course and learn how to apply solar technology to your home*

By Evan Powell

Solar energy is like the weather in that everybody talks about it, but few do anything about it. Perhaps it's a lack of fundamental knowledge, or exaggerated notions of required initial outlays, or the inevitable horror stories about troublesome systems—whatever the reason, there has been precious little action on what almost everybody admits is an appealing idea: getting "free natural heat" via the solar route.

There is nothing new about controlling solar energy. The pueblos of the American Indians of the southwest stored the sun's heat during the day and released it slowly during the cold, desert night and, in the 1930's, solar water heaters were used extensively in the South. Despite such scattered examples, solar energy never gained wide acceptance. While natural gas, oil and electricity were cheap and apparently abundant, we took year-round comfort for granted in our energy-wasting buildings.

Almost suddenly, the picture drastically changed. We know now that our energy supplies are limited and expensive. And we know that utilization of alternative choices, coal or nuclear energy, is a slow process and poses environmental questions.

Solar energy may be the ray of hope in this dark picture. On a clear day, almost regardless of season and temperature, the sun delivers about 250 Btu's per square foot per hour to the earth's surface. It is well within the realm of today's technology to build new homes using only this energy from the sun for heat.

But most of us live in yesterday's houses. While input from the sun would be a welcome supplement to our heating needs, applying solar technology to existing dwellings is often difficult and, all too frequently, based only on guesswork.

But you have a big advantage. No one knows your home and the needs of your household better than you. It is the aim of this primer on solar energy to give you what you need to know about selecting, sizing and even building equipment that will enable you to apply solar technology to your home.

COLLECTORS

CHIMNEY

DUCT TO LIVING AREA

PLENUM

STORAGE TANK     COIL PUMP     HOT WATER COIL

COLLECTOR CIRCULATION PUMP     FURNACE (AUXILIARY HEAT) AND BLOWER

**In active system** shown, when collectors' liquid is at proper temperature, sensor energizes collector pump to move liquid into storage tank. As heat is needed, coil pump forces liquid through hot-water coil in furnace plenum. Air is warmed as it is forced over coil by blower and is distributed by ducts.

254

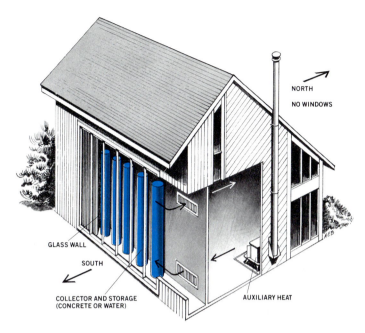

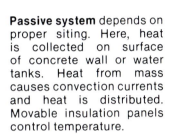

**Passive system** depends on proper siting. Here, heat is collected on surface of concrete wall or water tanks. Heat from mass causes convection currents and heat is distributed. Movable insulation panels control temperature.

NORTH

NO WINDOWS

GLASS WALL

SOUTH

COLLECTOR AND STORAGE
(CONCRETE OR WATER)

AUXILIARY HEAT

Gathering energy from the sun and putting it to use is a big challenge. How the challenge is met is usually determined by the application.

For residences, primary applications are space and water heating—ideal uses, not only because these are the No. 1 and No. 2 energy-users in the home, but also because solar energy can be applied directly to these purposes. It is also technically possible to use solar energy to generate electricity and to cool a house; however, with present technology, these applications are too costly for most homes.

Putting solar heat to work involves four factors: Collection, transfer, storage and distribution. In operation, solar radiation is absorbed by the collector and then transferred (normally, by air or liquid) to an insulated storage area with little heat loss. When heat is required in the living space, it is taken from the collector or, if none is available there, from storage. Also needed are a control system and an auxiliary heating system.

When fans, pumps or other mechanical systems are used to carry heat from the collector to storage, and from storage to the living area, the system is called *active*.

*Passive* systems are designed to use natural means to collect, store and distribute solar heat. For example, large expanses of glass on a south-facing wall may be used to collect heat, while a thick wall immediately behind the glass provides for storage. Or heat may be stored on the roof in large containers of water that are covered with insulated panels at night. Movable wall panels or flaps may be used to direct the heat, which can be distributed through the home by conduction, convection and radiation.

Passive systems, which require only a minimum of controls and ordinary materials, are relatively simple and long-lasting; however, since siting and design of the home are factors in determining the feasibility of this type of system, the technology is primarily used in new construction. There are some instances where a *hybrid* system—one that combines features of both active and passive systems—works well. However, for retrofit on existing homes, active systems are almost always best.

There are no absolutes in the world of solar energy. What works in one instance may not be right in another. But I'm convinced there is hardly any home that can't benefit from the application of solar energy in some form.

This chapter will acquaint you with some common types of solar systems—those that have proved effective. It will also go into some "tailored" methods for collecting and storing solar energy.

For example, a small greenhouse attached to the southern exposure of your home may be all you want and need. It would provide your family with plants and vegetables during the winter and help heat your home.

At the other end of the scale, one of the best applications I have seen of solar heating is an earth-insulated (partly underground) home developed by the US Department of Agriculture and Clemson University. A day and a night spent there gave me a real taste of solar living. During the night, when outdoor temperatures reached 16 degrees and winds gusted above 20 mph, the only sound was the quiet whisper of a small blower in the storage area below. During the period I was there, the house consumed only 1.1 kwh of electricity—in that area, less than six cents worth—and a good part of that consumption could probably be traced to the intermittent use of photoflood lamps for the pictures we were taking.

While it's unlikely that your home would benefit to the same extent from a solar system, even a fraction of that kind of saving could make a significant difference in your utility bills. And any difference that can pay for the necessary equipment within a few years is worthwhile.

# Solar collectors

*Their job is to trap heat so it can be used or stored*

## By Evan Powell

While solar energy is free, harnessing it for use is not. Any fuel savings realized in home applications can very well be offset by the initial cost of solar equipment.

Still, almost any home can economically utilize solar energy for one or more functions now dependent on other forms of energy. But the homeowner must (a) select the appropriate application for his home and (b) select the right equipment for the job. It is the aim of this chapter to provide the solar-energy basics you need to make such decisions wisely.

While significant strides have been made in developing photovoltaic materials that convert sunlight directly into electricity, such electricity is still a long way from being cost-competitive with what's available. Hot-water and space heating remain the most practical solar-energy applications for the home. And, for these purposes, good solar equipment, properly applied and installed, can deliver efficient, low-maintenance operation.

The amount of solar energy reaching earth is almost immeasurable. It takes only about 12 days for enough to arrive to equal all the earth's known fossil-fuel reserves. The Department of Energy estimates that the solar energy that reaches just 1/500 of the country—an area smaller than Massachusetts—could, if converted at only 20 percent efficiency, satisfy all our present electric power needs.

Despite this abundance, there are problems with the solar energy supply.

1. *Solar input is intermittent.* It is available only by day, and only when the sky is clear. As a result, it's necessary to be able to store it for those periods when it is not available.

2. *The sun's available energy is spread out very thinly.* The average maximum potential on a clear day is estimated at about 257 Btu per square foot per hour; when allowances are made for night and day, winter and summer, and cloudy and clear conditions, this average drops to about 58.5 Btu. Still, when this diffused energy is collected over a large area, it represents a significant amount of heat.

These factors make the collector perhaps the most important component of a solar system. Its purpose: To trap solar heat so it can either be transferred directly to use, or stored until it is needed.

257

## COLLECTOR PRINCIPLES

While it seems very different from solar collectors you see on buildings, the familiar greenhouse is a good example of a proven solar collector. It is based on the same principle as the flat-plate solar collector, the most common type in use.

In the case of a greenhouse, the visible, shorter wavelength portions of sunlight enter through the glass. When they are absorbed by the plants, ground and fixtures, they are changed to visible and longer heat waves, which are retained by the glass and so raise the interior temperature. The principle operates on cloudy, as well as sunny days.

If you study the flat-plate collector cutaway photo, you'll see how like a greenhouse this collector works. Visible light passes through the cover plate. The absorber plate prevents the light from escaping, so it heats the transfer medium— the air or water surrounding the plate. The medium then carries the heat off to storage or into the living space of the home.

## COLLECTOR CONSTRUCTION

Various components can be seen in the cutaway.

The cover plate usually consists of one or more layers of glass or plastic that are separated from the absorber plate to prevent re-radiation and to create an air space that traps heat by reducing convective losses. In some types of absorbers, air is evacuated from this space to enhance efficiency.

In "air-to-air" systems (air is the transfer medium), the transfer medium passageway may be nothing more than space between the cover plate and absorber plate. In a liquid system like the one shown, the passageway is usually a series of tubes that have good thermal bond with, or are an integral part of, the absorber plate. In some air-to-air systems, corrugated or deformed sheets may be used, with tubes connected by header tubes at the inlet and outlet.

The absorber plate is usually metal, with a dark surface coating. It should

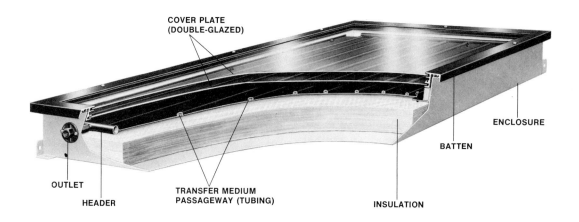

COVER PLATE (DOUBLE-GLAZED)

ENCLOSURE

BATTEN

OUTLET

HEADER

TRANSFER MEDIUM PASSAGEWAY (TUBING)

INSULATION

**Home designed** by Solar Fundamentals, Inc. of Greenville, South Carolina 29609, pictured here, is an excellent example of integral collector utilization. Unique setup is based on air-to-air collector system, but heat is transferred to water for storage by heat-pipe principle heat exchanger. Result is a high-efficiency installation that takes advantage of both air systems (low maintenance, no freeze-ups) and water systems (minimal storage space requirement).

offer as much surface area to solar radiation as possible. Insulation is placed behind the absorber plate to reduce heat loss through the back of the collector. The insulation must be suitable for the big temperature rise that may occur under dry-plate conditions and when no transfer-medium flow is taking place.

The enclosure contains all the components. It should be weather-tight.

What can such a system deliver? In a water-heating application, temperatures of 100 to 200 degrees F—enough to satisfy needs of almost any homeowner—are not uncommon. With a big enough collector area, it could also supply a large portion of the space-heating requirement.

Many other types of solar collectors are available. One, for example, is the concentrating collector, which focuses solar energy into a small area, often a single pipe that passes across the face. It works in much the same way as a magnifying glass used to start a fire. The concentrator can provide higher temperature ranges than the flat-plate collector, but it requires a tracking mechanism to insure that the highly reflective surfaces are always accurately aimed at the sun.

## ORIENTATION AND TILT

While positioning of a flat-plate collector is not as critical as for some other types, it must be oriented and tilted properly for maximum efficiency. The optimum tilt is usually the same angle as that of the site latitude. While it's a good idea to follow manufacturer's instructions, variations of up to 10 degrees seem to make little difference in performance of most collectors. There is always the possibility, too, that a deviation that takes advantage of radiation from nearby reflective surfaces will compensate for loss in direct solar energy. And, in many areas, the collector tilt may have to be modified to prevent snow from building up on the collector or drifting in front of it.

Orientation of the collector should be within 20 degrees to each side of true south. Again, though, allowances should be made for local conditions. Before locating collectors on your home, spend at least a day observing characteristics of sun and shadow. Obviously, a collector should not be in shade for long and, often, just locating it away from the chimney or roof overhang helps.

## MOUNTINGS

There are, in general, four ways to mount collectors (see drawings):

1.  *On racks.* Here, the structural frame may be on the ground or attached to the building or rooftop. It must be adequate to resist wind or other impact loads.

2.  *Standoffs.* A variation of rack mounting often used on finished roof surfaces, this involves placing the collector on standoffs, lengths of which can be adjusted to provide an angle other than that of the roof. Like the rack, the open standoff setup allows rain and air to pass under the collector.

3.  *Direct mounting.* Here, the collectors are attached right to the roof, usually over a waterproof film or membrane above the roof sheathing. (Weatherproof seal between collector and roof prevents leakage, mildew and rotting.) The roof's finish and flashing are then installed around the collector.

4.  *Integral mounting.* In integral mounting—often employed when a collector is designed into plans for new construction—the collector is attached to, and supported by the structural framing of the roof itself, and serves as the finished roof surface. Again, weather sealing is all-important to prevent structural problems.

Results possible with a collector built of ordinary materials that are readily available at the local builder's supply outlet can be amazing. You can also select from an array of components made by manufacturers. You can buy a matched system and install it yourself, or you can contract with a solar-heating specialist for a turn-key job. But, in any case, if you know enough about solar-energy to understand what you are getting, you're on your way to a problem-free system.

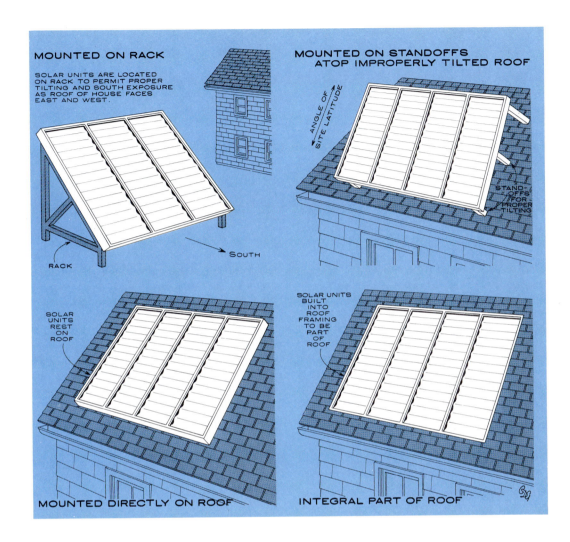

## MOUNTED ON RACK

SOLAR UNITS ARE LOCATED ON RACK TO PERMIT PROPER TILTING AND SOUTH EXPOSURE AS ROOF OF HOUSE FACES EAST AND WEST.

RACK

SOUTH

## MOUNTED ON STANDOFFS ATOP IMPROPERLY TILTED ROOF

ANGLE OF SITE LATITUDE

STAND-OFFS FOR PROPER TILTING

SOLAR UNITS REST ON ROOF

## MOUNTED DIRECTLY ON ROOF

SOLAR UNITS BUILT INTO ROOF FRAMING TO BE PART OF ROOF

## INTEGRAL PART OF ROOF

# How solar heat is stored

## By Evan Powell

On a clear, cold day, some of the heat captured by a solar collector might immediately be transferred from the collection area into the home. But a good collector system holds much more heat than what's needed immediately—on moderate days, there will be lots left over. Storing that heat for use at night, when the temperature drops, is the important job of the storage subsystem.

While there are many promising contenders, water and rocks or masonry are—and have always been—the favorite storage media. However, there are many refinements in the use of these media, and the lessons of the past can help you plan a good system. The information presented here relates primarily to space heating.

Usually, choice of a storage medium is determined by the collection and transfer medium. Active systems using air-to-air collectors normally use rock storage. Air-to-water collectors use water storage. Passive systems normally use masonry, though water may be used.

The amount of storage needed depends on your collector area and the material used as the storage medium. U.S. Department of Housing and Urban Development (HUD) minimum standards require stor-

**Passive system** storage is relatively simple. Brick wall behind glass at left has openings at top and bottom to allow convection currents to flow into living area. Movable shutters can mask off glass area at night or during summer. Water, usually in tall cylinders immediately behind glass panels, can also make excellent storage medium.— *Photo courtesy Brick Institute of America.*

age capacities of 500 Btus per square foot of collector area for an active system and 1,000 Btus per square foot for passive systems. One popular rule of thumb—unfortunately, not a very cost-efficient one—states that the ideal collector area for space heating is all you can get on the roof; the ideal storage volume, all you can get in the basement.

## ROCK STORAGE

Storage systems must be well insulated and tightly sealed. If rock is used, there should be some means of access to the storage area, since there's some possibility of problems—odors and dust and grease contamination from the recirculated air. In severe cases, such contamination could result in mold and mildew in the storage area. These problems can be largely overcome, however, by using smooth, clean pebbles and high-efficiency filters in the supply duct.

Rock storage systems should be designed so temperatures stay as even as possible throughout the storage area. Selection of materials plays a part in this, but inlet and outlet locations are also important. Normally it's advantageous to distribute air to the living space from the hottest section—near the top of the storage area. Return air should enter at the lower section of the storage area, preferably at the opposite end from the outlet, and be distributed along the bottom through a perforated duct or plenum chamber.

During collection, air should flow from the coolest part of storage to the collector (the cool air flow helps maintain collector efficiency). This is often accomplished by automatic or motorized dampers, which reverse the direction of air flow for collection and distribution modes.

## LIQUID STORAGE

The same principles apply to a hot water (or other transfer liquid) storage system.

Usually, heat from hot water is distributed by a direct system. After it is heated in the collectors, water is pumped into the insulated storage tank. When a room thermostat senses a temperature drop, it turns on a pump, which circulates water from the storage tank through a heat-exchange coil in the main supply duct of the warm-air system. Air is forced across this coil and into the living area by a blower (convectors or radiators are sometimes used to distribute heat).

With any kind of liquid storage, it's important to prevent "short circuiting" of water flowing straight from the inlet to outlet of the tank. To do this—and, again, to take advantage of the hotter temperatures near the top of the tank—simply locate the outlet near the hottest water at the top of the tank and the inlet for the cooler water at the bottom. Water storage generally requires substantially less physical space than rock.

## NEW DEVELOPMENTS

New methods for storing solar thermal energy are constantly being researched. A breakthrough in this area could mean a big jump ahead for solar energy usage.

One of the most interesting new storage systems involves eutectic salts. The most commonly used in Glauber's salt, or sodium sulfate decahydrate.

The salt systems work on the principle of latent heat and phase change in a material. You can understand how this works if you consider that it takes one Btu to raise one pound of water one degree Fahrenheit, but it takes 144 Btus to change 32-degree ice to 32-degree water.

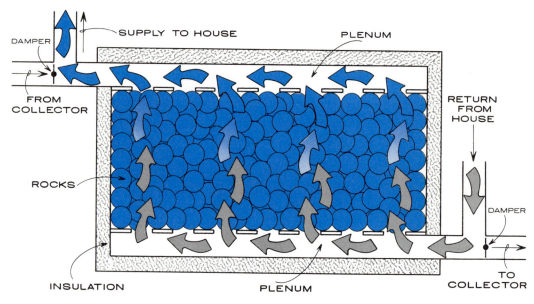

DAMPER
SUPPLY TO HOUSE
PLENUM
FROM COLLECTOR
ROCKS
INSULATION
PLENUM
RETURN FROM HOUSE
DAMPER
TO COLLECTOR

**Rock storage** area for warm-air distribution system (top) should be insulated and tightly sealed. Use of smooth, clean pebbles helps reduce maintenance. Plenum across top and bottom helps keep heat evenly distributed. Variations include water collectors with tank buried in rock pile. Convection from tank heats rock; air passing across rock flows into home area. Water or other liquid in a non-pressurized water system can be pumped directly from storage into heat-exchange coil or other distribution equipment (center). Outlet should be at top of tank, where hottest water is; return, near the bottom. In pressurized supply system, or whenever it's desirable to keep working fluid apart from storage fluid, a heat-exchanger coil can be used in tank (bottom). This "closed loop" setup is often used in hot-water systems to prevent air locks from forming in radiators.

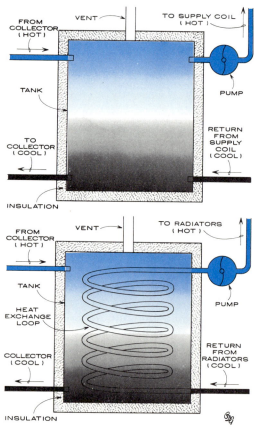

FROM COLLECTOR (HOT)
VENT
TO SUPPLY COIL (HOT)
TANK
PUMP
TO COLLECTOR (COOL)
RETURN FROM SUPPLY COIL (COOL)
INSULATION

FROM COLLECTOR (HOT)
VENT
TO RADIATORS (HOT)
TANK
HEAT EXCHANGE LOOP
PUMP
COLLECTOR (COOL)
RETURN FROM RADIATORS (COOL)
INSULATION

You can't measure this heat—it's absorbed or given off in a process other than temperature change.

That's what happens with eutectic salts. Glauber's salt is ideal for the purpose, since it changes from crystalline to liquid form at 89 degrees F, an ideal range for solar heat's typical low temperature levels. The latent heat available at this point gives additional capacity, which means less area is needed for storage.

But a problem can arise: Apparently, the salt tends to encapsulate and insulate itself, thereby reducing its effectiveness. New methods of storage and/or new salt formulations may help overcome this problem.

# The distribution system

*Whether air or liquid, the transfer medium must be "pumped" from one section to another*

## By Evan Powell

The collectors and storage facility are no doubt the most important features of any solar-heating system. Still, no matter how much heat your collectors gather, or how much you can store, the system serves no good purpose until the heat can be moved to where it's needed. How you move that heat depends in large measure on the type of solar system, and whether it uses air or liquid as the transfer medium.

To understand how the heat-distribution system works, think of it in the simple terms of a pump moving heat from one area to another. That "pump" may not actually be a pump in your system—if air is the transfer medium, it could be a fan, or in a passive system, convection currents. No matter—the "pump" is still moving heat and, to do this effectively, it must operate within a closed circuit so there's always an available supply of the transfer medium and an outlet. The diagrams on page 268 show what I mean.

There are other devices in the distribution system, also. For one thing, there will be conductors to carry the heat—ducts, piping, or simply openings in walls or passageways. And there may be diverters in the form of motorized dampers in ducts or valves in water lines.

Sizing of the components is important. If they aren't able to carry the volume of transfer medium that the rest of your system is capable of handling, they will tend to throttle the operation and reduce the system's effectiveness. Also, all runs —whether pipe or ducts—should be kept as short as possible, with as few bends as possible, since every inch and turn increases both the resistance to flow and heat loss.

You have probably adjusted dampers of a conventional heating system to control air flow into a room. Dampers in a solar heating system are the same, except they are motorized—the ducts open and close on command from a control circuit. The dampers may be designed to open and close completely, to stop in one of the several positions or to modulate as needed.

When a system uses a liquid transfer medium, the fluid flow may be controlled by solenoid-operated valves, or simply by the starting and stopping of the fluid-circulating pump. Many liquid-transfer systems have two or more loops, each with a separate circulating pump.

**"Hybrid" home,** built by Rural Housing Research Unit of USDA in cooperation with Clemson University, utilizes several technologies for maximum energy efficiency. Home is earth-insulated (underground) on three sides—only roof and front side are exposed. Front portion of roofline is air-type collector array.

An active air-to-air system must have a blower—usually, a squirrel-cage type connected to a split-phase motor. Normally, only one blower is used in a system; any required air-flow reversal is accomplished by the opening and closing of dampers.

There are also "hybrids"—systems that use heat exchangers to transfer heat from one medium to another, or from one circuit to another. The heat exchanger may be a coil of pipe placed in the air flow from a collector—hot air passing across it warms the pipe to preheat the domestic water supply. Or it may be a coil of pipe in the primary loop of a water-storage tank (see illustration, next page); often, the loop contains a transfer medium, such as ethylene glycol, which requires no special care when it is below freezing outdoors. A reverse application is a hot-water coil—often in a warm air duct—with air distributed into the home after it's been heated by collectors.

While many aspects of solar technology are extremely complex, the drawings should help you understand the basic distribution system.

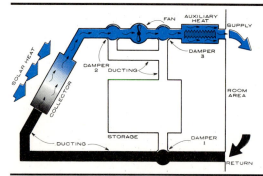

## Collector to living area

**Follow this** and the following two drawings, to see how solar heat moves, and is stored, in an air-to-air distribution system. In top drawing, living area requires heat, which is available in solar collector. Note damper position. Follow air flow to see how return air flows directly into collector, where it's heated and passed back into living area.

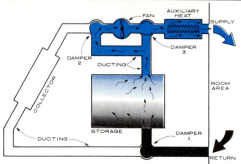

## Storage to living area

**Here,** living area requires heat, but none is available from collector (night). Air flow bypasses collector and flows through storage for heat. Note position of dampers. (Here, and in situation shown above, auxiliary heat can be added.) Temperature sensor in collector controls damper 2; others are controlled by room thermostats.

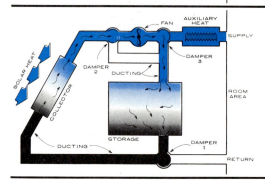

## Collector to storage

**In the situation** illustrated here, no heat is required in the living area of the home, but the collector is gathering solar heat. When the collector temperature gets higher than the storage temperature, air is circulated from the coolest point near the bottom of the storage area, through the collector, and then back into the storage area.

**In fluid-transfer system,** working fluid from collector is circulated through primary loop when sensors indicate that water in collector is heated. Secondary loop is engaged when control in living area calls for heat. Circuits may or may not be operating at the same time. When water is used as transfer fluid in the primary loop, freezeups may be prevented by circulating warm water through the collector, adding heat to the collector, or by the "drain-down" method—automatically draining water from outside system.

LEGEND

- COOL LIQUID
- HEATED LIQUID
- COOL AIR
- HEATED AIR

**How fluid transfer system works**

# Solar salts—
# New chemical systems
# store the sun's heat

*Clever ways to package phase-change salts may enhance solar heat storage*

## By Richard Stepler

For more than 30 years, researchers have tried to tame phase-changing materials for use as energy-storage media. Why? A given quantity of these sticky, corrosive salts can absorb and release far more heat than tanks of water or bins of rock, the usual ways to store solar heat for use when the sun's not shining (see charts, page 272).

Now, three new developments promise to make the use of such salts practical. All employ improved techniques to package the salt; and all involve improvements in the chemical makeup of the material.

- Sausage-shape packages of a patented, hydrated salt have been developed by the University of Delaware's Institute of Energy Conversion (photo, page 270). The salt melts and freezes at 55 degrees F and would be used in air-conditioning systems to store cool air

generated at night at off-peak rates. A variation of this system that would store solar heat during the day is also being developed by the Institute. The salt in this system would have a higher melting/freezing point.

- High-density polyurethane trays, filled with Glauber's salt and sealed using a spin-welding process (applying plastic to plastic under heat), are now being made by Valmont Industries (drawing on page 272).

- Also on the market is a six-foot-long, 3½-inch-diameter polyethylene tube filled with calcium chloride hexahydrate, which changes phase at 81°F. The thermal-storage compound was developed by Dow Chemical, with Department of Energy support. The Thermol 81 tubes are made by PSI, Inc., and cost $29.90 each. PSI estimates that about 100 of the tubes would be needed in a typical residential solar space-heating system.

Past problems with phase-change materials have prevented their use. These have included:

- Supercooling, the tendency of the salts to go well below their freezing point before beginning to solidify.

**Salt sausages:** University of Delaware borrowed technology from food-processing industry to develop low-cost polyester package.

• Encapsulation, the tendency of the material not to solidify completely. This reduces the salt's heat-storage capacity.

Dr. Paul Moses, Dow senior research chemist, explains that a key to the solution of these problems is adding a nucleator to the material to prevent supercooling. The nucleator starts the freezing process. "In addition," says Moses, "a vapor barrier is essential. If calcium chloride picks up water [it's hygroscopic], the salt loses its phase-changing ability rapidly."

Dow's energy-storage material is packaged in ultra-high-molecular-weight polyethylene tubes by PSI. The tube's end caps are thermally fused to permanently seal in the salt mixture. Thermol 81 rods carry a 10-year warranty.

Paul Popinchalk, an engineer with Valmont Industries, told me that Valmont has been working with Dr. Maria Telkes, a pioneer in phase-change materials. "We've changed the formulation of the Glauber's salt and the method of encapsulating it," he told me. "We're now replacing trays that didn't work in about 50 systems already installed." Valmont offers a five-year limited warranty.

Dr. Allen Barnett, director of the University of Delaware's Institute of Energy Conversion, says that solving the packaging problem is the key to making the system economical. Working with the Du Pont Company, and adapting food-packaging technology, the Institute developed a low-cost, sausage-shape package of multi-layered polyester film to contain its patented mixture of hydrated salts. The reliability and performance of Delaware's system are now being studied.

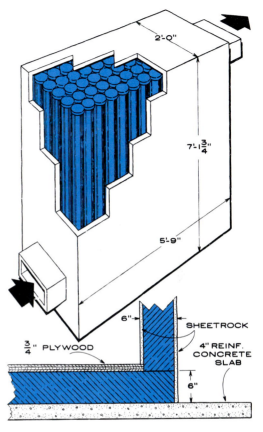

2'-0"

7'-1¾"

5'-9"

6"

¾" PLYWOOD

SHEETROCK

4" REINF.
CONCRETE
SLAB

6"

**Phase-change salt** is sealed inside PSI's 6'-long, 3½"-diameter polyethylene tubes (photo). Drawing (top right) shows typical storage container for an average home. Drawings at right show how tubes could be used in passive/active solar applications.

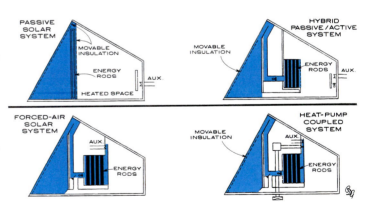

PASSIVE SOLAR SYSTEM

MOVABLE INSULATION

ENERGY RODS

AUX.

HEATED SPACE

HYBRID PASSIVE/ACTIVE SYSTEM

MOVABLE INSULATION

ENERGY RODS

AUX.

FORCED-AIR SOLAR SYSTEM

AUX.

ENERGY RODS

HEAT-PUMP COUPLED SYSTEM

MOVABLE INSULATION

AUX.

ENERGY RODS

## Storage materials compared

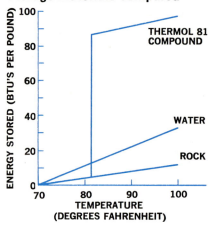

ENERGY STORED (BTU'S PER POUND)

THERMOL 81 COMPOUND

WATER

ROCK

TEMPERATURE (DEGREES FAHRENHEIT)

**Needed to store 300,000 Btu.**

| | Wt. (lb.) | Vol. (ft³) | Vol. (gal.) |
|---|---|---|---|
| Rocks | 37,000 | 480 | 3600 |
| Water | 12,000 | 380 | 2900 |
| Thermol 81 rods | 3700 | 78 | 600 |

## Solar heating with phase-change-salt heat storage

**Insulated cabinet** about 2′ wide, 2½′ high, 10′ long holds Glauber's salt in Valmont Industries' solar heating system. High-density polyurethane trays hold salt, which is sealed in "spin-weld" process.

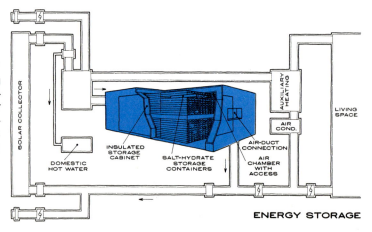

SOLAR COLLECTOR

AUXILIARY HEATING

LIVING SPACE

AIR COND.

DOMESTIC HOT WATER

INSULATED STORAGE CABINET

SALT-HYDRATE STORAGE CONTAINERS

AIR-DUCT CONNECTION

AIR CHAMBER WITH ACCESS

**ENERGY STORAGE**

# Solar-heat storage system solves an old problem in a clever new way

*Phase-change salts with a high melting point, tricky plumbing, and direct heat transfer, are the keys*

## By David Scott

Egon Helshoj picked up a shapeless, football-sized lump of salt crystal. "This is sodium thiosulfate pentahydrate," he told me. "We call it hypo for short. It's the basic chemical in our fast-acting heat-storage system for solar panels."

Like the better-known Glauber's salt, hypo is a phase-change material, absorbing large amounts of latent heat as it melts and releasing that heat as it solidifies. Unit for unit, materials that change phase can store much more heat than can water (maintained as a liquid) or a rock bed, the two most common-solar-heat-storage systems. That usually means both a dollar and a space savings.

Helshoj is managing director of Effex Innovation in Copenhagen, Denmark, and had brought his new solar-heat-storage system to the International Solar Energy Exhibition in Brighton, England. "This small one," he said, pointing to a five-foot-high cylinder, "could provide all the hot water for a house for three or four sunless days.

"Hypo has ideal characteristics for solar-heat storage," Helshoj continued. "Its thermal-storage capacity is exceptional, and it melts at a higher temperature than Glauber's salt: 48 degrees C (118 degrees F)—high enough to provide hot water and space heating, yet not too high for normal collector panels to tolerate.

"Until now no one has used this salt because of some undesirable physical changes that occur during the melt/freeze cycle," he explained. "But we've solved that in a simple way."

The Effex rig combines a heat-exchanger coil in the solar-panel and domestic-plumbing circuit with a tank of phase-change salts. The usual snag in such a system is that as the salts gradually solidify when latent heat is released, the crystals cling to the surface of the coil, building up an insulating crust. This soon stops further extraction of heat.

While there are mechanical solutions to the problem Helshoj has come up with an elegant, nonmechanical answer. He uses oil as a two-way heat-transfer medi-

273

# How it works

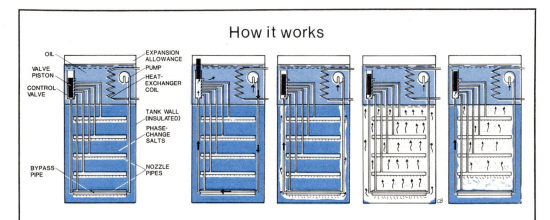

Heat-storage unit has five nozzle pipes in salt tank. When tank is fully discharged (first diagram, left) salt crystals block all nozzles. Second diagram: Charging begins when hot water, above 48 deg. C, from solar collector circulates through heat-exchanger coil, warming thermal oil. Electric pump starts, pumping oil via bypass pipe (inside lowest nozzle pipe) up to simple automatic distribution valve. Pressure lifts piston to its highest position, opening top outlet to circulate oil. Third diagram: Heat from bypass pipe melts crystals blocking bottom nozzles, opening oil passage from lowest valve outlet. The resultant pressure drop makes the piston fall to that level. Oil flow through the cleared nozzles causes further local melting. Crystals around vertical feed pipes also dissolve. Fourth diagram: Soon all the salt melts and the storage tank is fully charged. During this time the valve piston stays in the same posi-

tion. Discharge occurs when a cooler liquid, such as from home radiators at start-up, circulates through the heat exchanger. Oil chilled to below 48 degrees C initially circulates, causing tiny crystals to form in the solution. They rapidly increase in number and size, releasing latent heat, though the temperature remains constant. The heat is picked up by the oil and transferred to the liquid in the heat exchanger. Fifth diagram: Heavy crystalline slush sinks to the bottom of the tank and solidifies, first blocking nozzles in the lowest pipe. As pumping continues, back pressure in the oil-control valve lifts the piston to the next higher outlet, opening an oil passage to the second nozzle pipe. With further cooling a solid mass builds up progressively blocking the higher nozzles. At each stage, the piston rises until full discharge is reached. The pump then stops and the piston drops.

um between coil and salts. The special mineral oil, an immiscible fluid, is pumped through the molten salt.

Since the oil's specific gravity is lower than the salts', the oil rises to the top of the tank where the exchanger coil is immersed, and transfers the absorbed heat to it. A crafty multilevel arrangement for dispersing the oil through the salt solution beats the crusting snag. The drawings show how it works.

"This oil-feed arrangement plus direct salt-to-oil contact speeds the heat-transfer action," says Helshoj. "That means that fleeting bursts of solar warmth are captured, and even small reserves of heat can be extracted—and quickly."

After the Brighton exhibition, the hypo system was to be tested at the Danish Technical University.

# Heat-storage tube for passive-solar design

## By Jeffrey Milstein

Planning to use a phase-change material (PCM) in a solar heat-storage application?

PCM's are chemicals that store large amounts of heat by changing from a solid to a liquid. Two problems have plagued

PCM's: reduced effectiveness from chemical breakdown, and supercooling—stored heat isn't fully released at the melting-point temperature.

A new type of PCM tube from Boardman Energy Systems (Box 4299, Wilmington, Delaware 19807) solves these problems, says the manufacturer, and provides several melting points for various designs. The Boardman tube prevents

**Heat-storage wall** uses a bank of Boardman plated-steel PCM tubes. The tubes (below) are 4⅝ inches in diameter and 24 inches long, and weigh 23 pounds.

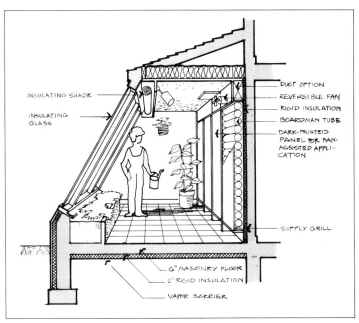

INSULATING SHADE

INSULATING GLASS

DUCT OPTION
REVERSIBLE FAN
RIGID INSULATION
BOARDMAN TUBE
DARK-PAINTED PANEL FOR FAN-ASSISTED APPLICATION

SUPPLY GRILL

6" MASONRY FLOOR
2" RIGID INSULATION
VAPOR BARRIER

breakdown and supercooling problems by fixing the PCM (sodium sulfate decahydrate or Glauber's salt) in a rigid matrix of cement that holds it in a linked cellular structure.

"Testing through over 2,000 cycles has shown no loss of efficiency," says inventor Brian Boardman. Previous designs lose efficiency, according to Boardman, when the melted salts drop to the bottom of the container. Repeated cycling gradually reduces the amount of water molecules available for combination with the salt.

A Boardman tube with an 89-degree-F melting/freezing temperature stores 2,100 Btu of latent heat per tube, says the inventor. A 67-degree tube—suitable for the lower temperatures of passive and passive-hybrid designs—stores 1,900 Btu. A 45-degree tube is suitable for air conditioning.

The firm says that about 40 tubes, requiring only two by three by five feet of space, are suitable for a well-insulated house with an active or passive-hybrid design.

# Reflective insulating blind traps solar heat

## By Edward Moslander

Although large south-facing windows are one requisite for a passive solar heating system, they don't assure the success of a passive installation. The biggest problem is nighttime heat loss through the windows—it can cost all that's been gained during the day. Also, furniture and carpets can become bleached and faded after constant exposure to the sun's rays; "hot spots" may develop in the front of a room where the sun's rays are brightest; and in summer, a room can overheat, raising air-conditioning costs.

These drawbacks have made some homeowners reluctant to install a solar heating system. But at Oak Ridge National Laboratory, Hanna Shapira and Randy Barnes have developed a system they claim eliminates window problems. The reflective insulating blind (RIB) both admits and distributes solar energy during the day and prevents excessive heat loss at night by using louvered blinds, similar in appearance to Venetian blinds.

The blind consists of 3.5-inch-wide slats with a curvature of 12-inch radius. The Oak Ridge models are wood, covered with aluminum foil. The curvature of the blind causes sunlight striking it to be reflected up to the ceiling, which acts as a

**View outdoors** is only slightly reduced by RIB, due to fewer and wider slats.

heat-storage medium and light diffuser. It then radiates down, heating the room. The curved louvers eliminate the need for making continual adjustments during the day to keep the light reflected upward.

What makes RIB special, though, is that it performs a second function. As

Hanna Shapira explains, "This is the first system that, in addition to admitting solar energy, also prevents heat from escaping at night through the windows. When the blinds are closed they form an insulating shutter. And additional insulation comes from the air gap between the window and the RIB unit. Heat loss is greatly reduced." The result: little additional energy is needed to heat the room the next morning.

Since light is reflected to the ceiling and radiates down, RIB overcomes several problems simultaneously. Light does not directly strike furniture or carpets, eliminating bleaching. Heat is not concentrated, so hot spots are not a problem. And light is directed to the center and back of the room, eliminating the problem of glare in the front and near darkness in remote corners.

The ability to alter RIB's angle of reflection keeps rooms from overheating in the summer. Reversing the direction in which the sun is reflected—outside instead of in—keeps the room cool and lets in light at the same time, thus decreasing air-conditioning costs.

A major problem in designing RIB was that it must cover the entire window to be effective. "We had to make sure the view would not be obstructed. We decided that with fewer, wider slats little visibility would be sacrificed," says Shapira. In fact, the view is less obstructed than it would be with conventional Venetian blinds.

Shapira and Barnes claim that, depending on location and climate, over 50 percent of the energy required during an average heating season could be saved after installing an RIB unit in a passive direct-gain system. At the Oak Ridge unit, even higher savings resulted. During an average heating season Randy Barnes measured consumption at 20,212 kWh. In an energy-efficient structure without the

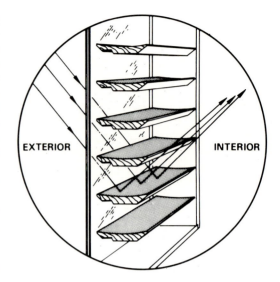

DAYTIME MODE

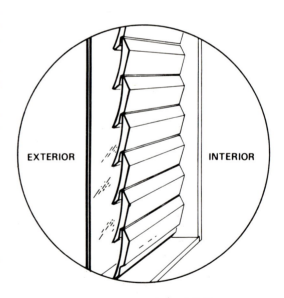

NIGHTTIME MODE

**During the day** RIB reflects light up and into room (top). At night slats close to form insulating shutter, trapping heat in.

RIB unit, that figure was reduced to 7,589. After installing an RIB system, only 3,779 kWh of energy were used—a reduction of nearly 81 percent. Most of the savings comes from the heating effect of incoming sunlight, the rest from reduced nighttime heat loss. With the spread of light throughout the room, additional savings can result from reduced dependence on artificial light. This was a bonus saving RIB's makers didn't really expect.

What is RIB's future? Shapira and Barnes designed RIB with residential use in mind, but there are possibilities for commercial use. Both options depend on what the search for materials turns up. "Right now, several manufacturers are exploring what materials might be used. RIB needs a material that's a good insulator with a reflective material on top," says Barnes. "We're waiting to see what's best."

# Air Floor stores and radiates heat

## By Jeffrey Milstein

Thinking about building a solar house? If so, you're probably considering the popular passive-hybrid system, and wondering how best to store excess heat. Air Floor might be just what you're looking for. It lets you store warm air in a concrete floor and provides comfortable radiant heat.

A single Air Floor pan is a 26-gauge steel stamping. It's 12 inches square and three inches high.

Here's how you use Air Floor: First, you pour a two-inch concrete base. Then you place the Air Floor pans on the base and connect them, covering the entire base. Finally, you pour the concrete floor over the Air Floor. In effect, you now have a hollow concrete floor into which warm air is fed continuously.

In commercial applications, warm air is forced into the Air Floor and exits through floor ventilators. About half the heat now comes from the floor outlets, and half is radiant.

In a passive-hybrid design, excess heat from a high point in the house is forced by a fan through ducting into the Air Floor. This keeps the high point of the house from overheating, while at the same time providing heat storage for sunless periods. The several inches of concrete poured over the Air Floor store the heat, radiating it out into the living area uniformly and comfortably.

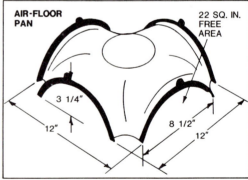

Another interesting possibility for Air Floor is summer cooling. By blowing cool night air through the floor, you could use it as a storage battery to cool your house during the day. This would be ideal for hot climates that have cool nights.

Air Floor is made by Air Control Systems, 14625 Carmenita Rd., Suite 220, Norwalk, California 90650. A floor pan will cost between $1 and $1.50, the company says, depending on where you buy.

# Translucent water wall stores heat

## By W.J. Hawkins

Pumps, tanks, heat exchangers, sensors, controls, plumbing, electrical wiring—they're all just some of the "extras" you must consider when installing a solar heating system in your home. But now there's an easier way: a passive "water wall."

Made by One Design (Mountain Falls Rte., Winchester, Virginia 22601), Waterwall is a four-foot-square translucent polyethylene module that mounts between the studs in a wall or roof. After installation behind a double-thick acrylic glazing material that you put on the outside of your house, you fill the module with water. The idea: Sunlight passing through the glazing warms the water, which retains the thermal energy (442 Btu/deg. F). When the sun sets, the modules release this heat into the home.

The interior wall can be covered with conventional materials. But since Waterwall is translucent, it allows sunlight into the room if left bare. The modules can be used in new construction or added to existing walls or roofs. Gray panels are about $110; white, $125.

**Waterwall modules** mount in south wall or roof. Weight is 31 pounds empty, 442 pounds filled with 53 gallons of water.

# Solar water heater—
# No pumps, no tanks

## By Ed Moslander

Heating water with solar energy isn't difficult or expensive: Just fill a bucket, put it in the sun, and wait. The money comes in when you want to tie your bucket (collector) into your household water supply. That usually means pumps, storage tanks, heat exchangers, thermostats, and a lot of plumbing—which costs $2,000–$3,000 when installed in your home.

A new solar water-heating system that dispenses with much of that hardware yet is claimed to perform about as well as a conventional system should be on the market in about a year. Solaron Corporation, makers of solar space- and water-heating systems, recently acquired the rights to market the simplified system developed by W. Peter Teagan of Arthur D. Little Inc., Cambridge, Massachusetts.

Solaron has not yet announced a price for the heater, but Teagan says the installed cost should be about 40 percent less than units now available. If so, the shortened payback period could make solar water heating attractive to many homeowners.

"I realized that the cost of solar water heaters wasn't going to come down without some fundamental changes," says Teagan. His solution: Connect the collector directly to an existing hot-water system and use city water pressure for circulation. Result: The device needs no solar storage tank, circulation pump, or differential controller, and uses less piping.

The collector consists of an absorber tank containing water and a heat exchanger. An intake pipe brings water by line pressure through the exchanger and then into the domestic hot-water tank, where it may be further heated to supply adequate temperatures at the tap.

The mystery of the system is why the water in the absorber doesn't freeze or boil. It contains no anti-freeze, and it is in direct contact with outdoor conditions. Solaron claims that insulation around the sides of the tank and the large mass of water in it prevent freezing. To eliminate overheating, a finned tube on top of the collector passes excess heat to the air and keeps the water temperature just below the boiling point.

The unit was field-tested through three New England winters without any freezing or overheating, says Teagan. On a clear day, the temperature of the absorber climbs 20 to 25 degrees F an hour. Due to the storage capacity of the unit, preheated water can be delivered well after sundown.

Because the storage is not isolated from ambient temperatures, collector performance is somewhat worse than with conventional units in cold months, but it may be somewhat better during hot months. How much of your water-heating needs could it supply year-round? Up to 50 percent, says the inventor. Solaron (1885 W. Dartmouth Ave., Englewood, Colorado 80110) will market the system for do-it-yourself installation.

# Two-tank system for solar-heated hot water

## By Susan Renner-Smith

What's so different about a two-tank system? Plenty—when the second tank acts as a reservoir for stored heat energy. In most solar hot-water systems, a second tank simply stores backup hot water for domestic use. But the hot water in this 150-gallon energy-storage tank (see diagram) never flows to your taps. Instead it circulates in a closed loop between tank and solar collectors.

Meanwhile, cold water from your house main flows through the heat-exchanger coils in the large sealed tank. The now-heated water flows back to your existing water heater (no need to replace it with a special tank as in many systems).

Since cold water never mixes with solar-heated water in the sealed tank, the water doesn't stratify and lose energy as it does in other tanks. Hefty R-14 foam insulation also helps retain the heat energy. The giant tank stores so much energy, claims the maker (Sealed Air Corporation, 3433 Arden Rd., Hayward, California 94545), that it will heat your hot water during several days of cloudy weather. Only after a long sunless spell will your conventional water heater kick in. In the

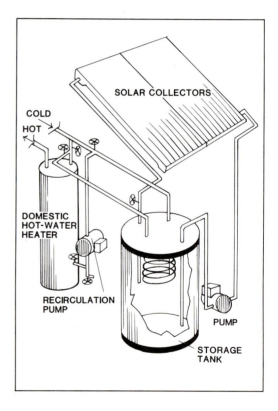

meantime it serves merely as a storage tank for domestic hot water.

The system's solar collectors are also unique, says the company. Each panel is one piece of extruded thermoplastic with oval water channels. The design permits the circulating water to contact more sun-heated surface than other collectors, it's claimed.

Sealed Air Corporation says that it has four years of in-use test results to back up its claims. It offers a five-year-warranty for the $2,500 system.

**A galvanized-steel frame** houses the one-piece thermoplastic collector panel (above). It's kept in place by guides at each end.

# Now: A do-it-yourself solar heating system

---

*It's efficient, low-priced—and the only tool you need is a sharp knife*

---

## By Evan Powell

A close look at your current water or home heating bills may convince you that now is the time to make some use of solar energy. You may even decide to save a bundle by doing the work yourself.

If you've checked into this possibility, you've probably been told that a solar collector, the biggest part of such a project, is beyond the scope of a do-it-yourselfer. A collector needs exotic, special coatings and proper welding or bonding. The cost of most collectors reflects this technology.

But there's good news. One of the most efficient solar collectors yet devised, and the first ever recommended to the Department of Energy by the National Bureau of Standards, is ideally suited for do-it-yourself installation—and it is also one of the most inexpensive collectors you can buy.

**SolaRoll** (facing page and detail drawing) is attached to roof or wall with adhesive, requires no nails, screws or penetrating fasteners. It's easy to use around vent stacks and other obstructions, making it a good choice even for very large collector areas.

**Drawing,** right, shows section of collector, giving detail of one loop. Mat is cut twice length of desired collector and doubled back to form two passes. Tubes connect to manifold to allow counterflow in alternate tubes throughout system to maximize heat transfer.

**Completed collector** might look like this, which uses ten passes, or five mat lengths. Note raised framing on lower edge; it forms flashing on collector edge.

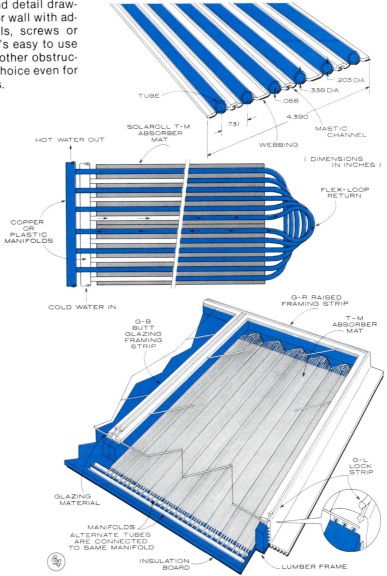

Known as SolaRoll, the system is based primarily on a six-tube mat material made of EPDM (ethylene-propylene-diene-monomer). It withstands temperatures of −60 to +375 degrees F. It's not subject to corrosion, ultraviolet light or chemical deterioration. And, since it's impregnated throughout with carbon black, there are no coatings to peel, chip or deteriorate. Its estimated life span: around 30 years.

SolaRoll comes in 600-foot rolls. The mat strip consists of six tubes (see illustration), separated by narrow strips of webbing that act as heat collection plates. (Note the channels in the underside of the webbing. They can be filled with mastic

**SolaRoll mat** is removed from 600-foot reel and cut to twice the length of the completed collector, seen at top, left, rolled up and ready for installation. Initial fabrication is done on the ground. At midpoint of desired collector length (top, right), webbing is cut away with sharp chisel or knife. Once webbing is removed (bottom, left), mat can be turned back to form two passes of collector (note crossover pattern of tubing). At the manifold end (bottom right), mat is cut and webbing again stripped from both ends. Then, every other tube is cut back to allow alternate tubes to be connected to the same manifold.

during installation to bond the collector to the roof wall without nails, screws or other penetrating fasteners.)

Also a part of the system are framing strips, made in several configurations, for holding glazing in position. Like the mat, the strips can be mounted with adhesive. A raised strip does double duty as flashing around the collector. When glazing is in place, a lock strip is inserted into a channel; it exerts constant pressure on the glazing to effect a good seal. It's pos-

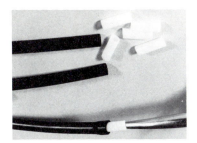  

**Using a lubricant** (left), unique Teflon jamb sleeve is inserted into end of each tube. After sleeves are inserted (center), tubes can be thumb-pressed into specially designed holes of the manifold. Watertight connection can withstand more than 60 psi pressure. With glass, Filon or other glazing material in place (right), a lock strip is placed in the channel to exert constant downward pressure to maintain seal. Framing strips are also made of EPDM, and they are attached with mastic.

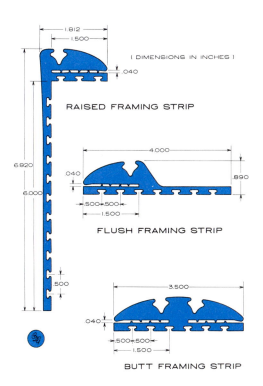

( DIMENSIONS IN INCHES )

RAISED FRAMING STRIP

FLUSH FRAMING STRIP

BUTT FRAMING STRIP

**Various kinds of framing** include: Raised—also acts as flashing; flush—used when glazing is flush with the wall or roof; and butt—used to support glazing between two adjoining sections of collector. Finished job (right) is neat and nearly maintenance-free. Versatility of strips makes SolaRoll suitable for solar walls, too, and for retaining glazing in greenhouses, skylights.

sible to install the material using no tool other than a sharp knife.

Because it is lightweight, the SolaRoll collector usually is built on the ground, and then rolled up and lifted into position for installation. The flexibility of the material allows you to build it to almost any useful size—up to 24x75 feet. The vertical manifold tubes run side by side at one end—rather than at each end—greatly simplifying plumbing hookups.

The accompanying illustrations give installation details. SolaRoll is manufactured by Bio-Energy Systems, Inc., Box 87, Ellenville, New York 12428.

# Solar electric home: The high-tech way to energy independence

By day this house sells electricity to the utility; at night it buys it back

## By Richard Stepler

When the sun shines, the family living in this large contemporary home will not be buying any power from the local utility. Instead, Boston Edison will be crediting their account for much of the electricity produced by the shiny panels for solar cells on the home's south-facing roof.

"In fact, the photovoltaic array produces so much electricity that there's a very good chance the house will be a net energy producer and exporter by the end of any given year," says Steven Strong, president of Solar Design Associates (Lincoln, Massachusetts), the engineering and design firm responsible for the home.

Sound too good to be true? Strong is quick to add a caveat: "Given the present cost of photovoltaic modules, the system is not economically viable. But then it's not meant to be; it is meant to be a demonstration of technical viability."

Indeed, with a cost of about $75,000 for the home's 7.3-kW photovoltaic system, economic justification is not even remotely possible.

The Carlisle home is only one facet of a Department of Energy program that aims to develop photovoltaics—the direct conversion of sunlight to electricity—for practical home use. Other prototype systems are operating nearby in Concord at DOE's Northeast Residential Experiment Station. Another experimental station is located in Las Cruces, where prototypes are being tested by the New Mexico Solar Energy Institute. A third station, to be located in the Southeast, is planned for the future.

Goal of the DOE program: off-the-shelf photovoltaic systems that cost $1.60 per watt by the end of 1986. (Today's solar cells cost about $7.50 per watt.) For the homeowner it means that a five-kW system—adequate for average homes—would cost about $10,000 installed.

At the Northeast Station, operated by MIT Lincoln Lab, systems designed by General Electric (5.8 kW), Westinghouse (5.4 kW), TriSolarCorp (4.8 kW), Solarex (6.2 kW), and MIT Lincoln Lab (6.9 kW) are now undergoing testing. The systems are "utility interactive," with the power produced by the photovoltaic arrays con-

**One thousand square feet** of photovoltaic panels on roof of 3,100-square-foot home produce 7.3 kW. The array's 9,500-kWh annual output will comprise 60 to 100 percent of the electricity the home uses annually. The house will sell a projected 1,300 kWh to the utility and buy 2,800 kWh a year. Equipment room (inset) holds monitoring gear, power-use meters.

verted to the 110/220-volt 60-Hz single-phase current supplied by utilities. During the day the arrays feed excess power into the utility grid.

"For the Carlisle house, only 25 to 28 percent of the photovoltaic output will be used for on-site needs," MIT's Miles Russell told me. "The rest goes back to the utility."

At night or during cloudy weather, the house draws power from the grid. There are no batteries for on-site power storage; if there's a power failure, the system shuts down until the utility comes back on line.

The system consists of the photovoltaic array in the form of plug- or wire-together solar-cell modules, a power-conditioning unit that converts the cells' direct current to alternating current, and control equipment.

## LIVE-IN EXPERIMENT

"In the first phase we build an unoccupied structure with a full-size photovoltaic system," Russell told me. "We perform tests to ensure safety, reliability, and durability. Then we go to lived-in situations."

The Carlisle house is the first such lived-in experiment. Built on speculation by Builders Collaborative, Inc. (Acton, Massachusetts), the house has now been sold. The buyer agreed to permit monitoring by MIT for up to five years. (He also keeps the photovoltaic system when monitoring is complete.)

The home's photovoltaic array consists of 126 two-by-four modules made by Solarex Corporation. Each module contains 72 semi-crystalline silicon cells and is rated at 58 W of peak power under full sunshine. Total rated daily output of the array is 7.3 kW.

The modules are wired so that 14 are connected in series to match the input voltage of a Windworks Gemini DC-to-AC inverter. This power conditioner matches the array's output to the utility's 60-Hz electric service.

There's a full complement of appliances inside the solar all-electric home: heat pump, dishwasher, clothes washer and dryer, vented range and oven, refrigerator, freezer, well-water pump, even a whirlpool bath. The home is frankly luxurious. According to Steve Strong, the home's size was largely dictated by the DOE-required 1,000-square-foot solar-cell array. It takes a large roof to hold that many panels.

While the home is large, it is also energy efficient, with a heat-loss coefficient of four Btu per degree-day per square foot—about 2½ times better than conventional construction. This reduces the load on the photovoltaic system. Architect Robert Osten of Solar Design Associates achieved the energy saving through a number of design features:

- Windows are few on the north, east, and west—and triple-glazed.
- The house is super-insulated with double, staggered 2 x 4 exterior walls (R-30).
- Ceilings have 12 inches of fiberglass (R-40).
- Rigid foam insulation is placed on the outside of the foundation.
- A three-foot-high earth berm shelters the first floor.

Passive solar heating plays a significant part in heating the home. The south wall contains some 350 square feet of floor-to-ceiling double-glazed sliding glass doors with heavy drapes to hold in the heat at night. Clerestory windows above the photovoltaic array provide additional solar gain in winter; in summer they can be opened for ventilation.

Heat is stored in a four-inch-thick, quarry-tile-covered concrete slab. A massive masonry fireplace in the living room provides additional heat-storage mass, auxiliary heat, plus a return air duct that

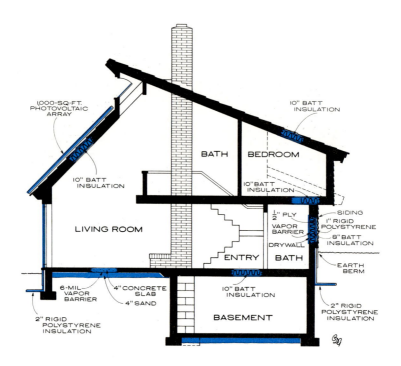

supplies preheated air to the heat pump—an air-to-air 38,000-Btu dual-compressor unit made by Carrier. If needed, it's also used for summer air conditioning.

Hot water, perhaps the home's second most energy-thirsty need after space heating, is largely handled by 100 square feet of thermal-type solar collectors (visible in the inset roof section in the photo). Backup water heating is provided by a heat-pump water heater.

What's the significance of a *lived-in* photovoltaic home? For one thing no one's sure how much energy a family will get from the solar cells.

Russell told me that MIT is also monitoring the electric-use habits of five nearby conventional homes. "Typically, loads appear early in the morning," explains Russell. "People wake up and turn on the lights and the *Today Show*.

There's a lull during the day because occupancy is low. Then everyone comes home in the evening, turns on lights and the TV, and starts cooking."

But the family living in the Carlisle house will be well aware of the photovoltaic system; they'll know it's in their interest to run appliances when the sun is shining. "You can do the cooking and baking, and run the dishwasher and the washer and dryer when photovoltaic output is high," says Russell. "You postpone them when it's low."

Of course, you can conserve only so much. "You tend to cook when you're hungry, and the refrigerator just marches along," concedes Russell. "And if they fill the house with electric gadgets, and have a typical bimodal peak in load—high use in the morning and early to mid evening—they'll use more utility-generated power."

## 3,100 square feet of living space on two levels

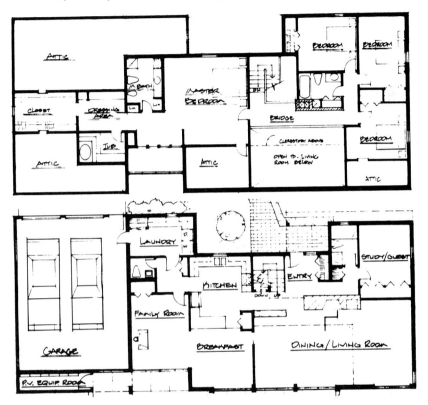

Russell thinks that the photovoltaic contribution for the Carlisle house will fall between 60 and 100 percent of the load, but that "somewhere in the middle is what we'll get—about 80 to 85 percent." Steve Strong is more optimistic, predicting that the homeowner could produce more energy than he uses "if he practices a certain amount of conservation and is not blatantly wasteful."

During the monitoring phase of the project, Russell and Burt Nichols, who's with MIT's Solar Photovoltaic Residential Project, will chart the progress of the Carlisle house. They'll measure array, inverter, and utility power flows, along with insolation, inside and outside temperatures, and other data.

"The hardware is here, the concepts are here; they're perfected, they work," says Strong. "The last barrier is cost." He believes that the cost goals of the DOE program will be met "given a mass-production market. The problem is mainly over-opportunity. As your article pointed out there are a lot of different options in photovoltaics, and the people who have the capital to invest aren't sure where to put it."

Another potential problem for home photovoltaic systems is how much the utilities will pay for the power generated. "The Federal Public Utility Regulatory Act mandates that they must buy the power," explains Strong, "but it does not mandate the price."

At least one state already has an attractive rate: New Hampshire's public-utility commission has set a price of 7.8 cents per kWh for power produced by on-site systems. "It's been challenged in court," says Strong, "but I suspect it has a fair chance of being upheld." But look before you leap into the power generation business. In most areas, the rate structures are still highly flexible, and most seem to be a combination of several structures, depending on the amount of power you intend to furnish. They also are influenced by time-of-day factors and demand. In many areas, you may also be required to furnish or pay for part of the necessary interconnecting equipment to insure that the power you furnish meets the standards required of the utility and does not influence other customers or personnel.

# Solar heating accessories—
# In brief

### Solar roof vent

Install its FP-10 roof vent and you'll help prevent blistering and cracking of built-up roofs, says Johns-Manville (1601 23 St., Denver, Colo. 80216). A one-way valve opens as sunlight expands moisture-laden air trapped in the roof. The valve closes when the roof cools.

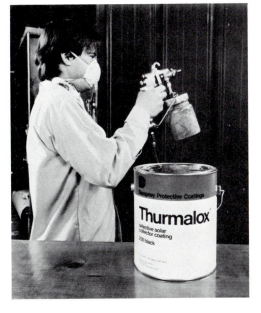

### Collector coating

Thurmalox black solar coating selectively absorbs solar wavelengths with the greatest energy, collecting heat more efficiently than ordinary black paints, claims Dampney Co. (85 Paris, Everett, Mass. 02149). The company says the paint resists outgassing and photodegradation by UV light.

## Flat-plate frame

The rear leg of the Sunrack frame telescopes so you can position your collectors at their optimum tilt, regardless of roof slope or orientation, says Sunsearch, Inc. (Box 275, Guilford, Conn. 06437). The aluminum frames are said to withstand 120-mph winds.

# 9 HIGH-EFFICIENCY WATER HEATERS

**H**eating the living space in your home is only part of its operating cost. Right behind it is the cost of energy used for heating water. This chapter tells you how to deliver a one-two punch and score a fast KO in the round with the water heater.

Any of the concepts offered here work best in conjunction with a program of water conservation in the home, especially hot water. Specially-designed water-saving showerheads can cut your present costs way down, even with your present water heater. Reducing water temperatures can help also, and reduce the danger of burns at the same time. But equipment counts for a lot, and the devices found in this chapter not only work well but pay for themselves quickly.

# Heat-pump water heaters—Best choice for your home?

*How much do you spend to heat water? Would you save with a heat pump? Or should you go solar?*

## By Evan Powell

Howard Patterson, an electrical contractor and general innovator, led me to the equipment room of his suburban Atlanta, Georgia, home to show me the latest development in heat-pump water heaters. "There it is," he said, pointing to a small, blue metal box. It looked about like all the other heat-pump water heaters I've seen.

But it doesn't work like the others. While they take heat from room air to heat the water, this one takes heat from well water to do the job. It's a water-source heat-pump water heater. The idea for this machine was Patterson's, and he persuaded E-Tech, an Atlanta firm that makes air-source models, to put it into a limited production run.

That's only one of the recent innovations in this fast-growing field. Now there are several versions of air-source

heat-pump water heaters as well, including both add-on models you attach to a conventional water heater (which then serves as a storage tank and backup heat source) and self-contained models—integrated units that combine tank and heat pump in one cabinet. About a half-dozen manufacturers now produce them, including one major air-conditioner maker, Fedders (see table at end of article). Two years ago only two small companies, E-Tech and Energy Utilization Systems, were in the business.

Would a heat-pump water heater be a good investment for you? If so, what kind should you buy? Or should you install a solar water heater instead? Here's what you need to know to decide.

## HEAT-PUMP BASICS

The reason heat-pump water heaters are proliferating is simple: They're very efficient. As with heat pumps used for space heating, under many conditions they use less electricity to provide a given amount of heat than does an electric-resistance heater, because they use standard refrigeration techniques to absorb some of the heat needed for the job from surrounding air or from water.

**Add-on heat-pump** water heater such as this Fedders is attached to a conventional water heater, which is used as a storage tank and for backup heat. The heat pump is a compact box that can be located as much as 200 feet away from the tank.

An electric-resistance heater uses one Btu of electric energy for every Btu of heat energy it produces. Engineers express that as a ratio called the coefficient of performance (COP). Thus the COP of resistance heat is one. I have two air-source heat-pump water heaters, and they both operate with an average annual COP of 2.3—meaning for every Btu of electric energy they use, they produce 2.3 Btu of heat energy.

Patterson's water-source heat pump does even better. He has thermometers and recording wattmeters monitoring his heater just as I do mine. As I checked the readouts the difference surprised me. His was delivering a COP well over 4.0. The reason for the greater efficiency is its source of heat—the well water, which remains at a relatively high and stable temperature (about 60 degrees) year-round. The air mine draws heat from fluctuates considerably, winter to summer. Even considering the electricity needed to draw water from the well, the water-source unit's COP averages more than 3.7.

A heat-pump water heater has generally slower recovery than a conventional electric heater with a 4,500-watt element; the Btu output of the E-Tech, for example, compares to a conventional 3,800-watt element. But it provides faster response: The flow-control valve provides hot water for instant use at the tank's outlet.

## THE ECONOMICS

According to the Department of Energy, water heating accounts for 16 percent of our national energy consumption and is the biggest user of electricity in the home (28 percent). The average family with an electric-resistance water heater, the most common kind, uses 6,600 kWh annually at a cost of $330.

Other types of water heaters do the job for less. Here's how they all compare, in cost per million Btu of heat energy, assuming average costs for fuel and typical efficiencies for heaters—100 percent for electric resistance, 80 percent for natural gas and propane, 70 percent for oil.

The cheapest form of readily available energy is natural gas, at least in its present regulated form. At 50 cents per therm it costs about $6.25 per million Btu. In second place is LP gas; at 80 cents a gallon the cost per million Btu is $10.87. Next is oil; at $1.25 a gallon the cost per million Btu is $13.13. Then comes electric resistance. At six cents per kWh, a million Btu cost $17.58.

## Would a heat-pump water heater save you money?

Make these calculations and you can determine how much you spend for water heating and how much you'd likely save with a heat pump. To do so you need a comparison factor for each fuel, which I determined by dividing the available Btu in that fuel into 1,000,000. To find your cost per million Btu with your present water heater and with a heat pump, multiply the appropriate factor given by the price you pay for a unit of that fuel.

For a natural-gas heater, the comparison factor is 12.5. (If gas costs you 50 cents per therm, for example, multiply that times 12.5 and you find that gas water heating costs $6.25 per million

Btu.) For LP gas use a comparison factor of 13.586; for fuel oil use 10.504. Multiply each by its cost per gallon. For electric-resistance heat, multiply your price per kWh of electricity by 293. To find what you'd pay with a heat-pump water heater use a comparison factor of 127.4 for an air-source unit and 73.27 for a water-source; multiply each by your cost per kWh. One million Btu will supply the average family of four with 120-deg.-F water (at an 80-deg. temperature rise) for about 15 days. Comparison factors also account for typical water-heater efficiencies (see text).

But use the electricity in a heat-pump water heater, and the saving is dramatic. An air-source version, realizing a year-round COP of 2.3 as mine does, ranks just higher than natural gas—$7.64 per million Btu. And the water-source version is the hands-down winner at only $4.39 per million Btu. (The box shows how to figure your cost and likely savings.)

The other side of the economic question is initial cost. A heat-pump water heater costs more to buy than other types. Conventional high-efficiency heaters cost from $200 to $400, plus installation. A heat pump will cost around $850 to $1,800 installed; $650 to $1,500 if you install it yourself. With factory-made kits, the job is well within the skills of most do-it-yourselfers.

Most manufacturers claim your operating savings will recoup the higher initial cost in four to five years, but that depends on many variables. In general, if you're now using natural gas for water heating, only a water-source heat pump is likely to be cheaper to operate. And since the air-source type removes heat from the

house, it may not be a good investment if you live in a very cold climate and have no source of waste heat it can draw from.

While the products are too new to have an established durability record, most studies indicate an expected lifetime of 12 to 15 years. Most companies offer a one-year warranty on parts and labor; Oregon offers a three-year and E-Tech a five-year warranty.

But before you decide to switch to a heat-pump water heater, you may want to consider going solar instead.

## THE SOLAR COMPARISON

The town of Lafayette, Indiana, was literally wrapped in darkness when Dennis O'Neal and I drove through the snow to dinner in February 1978. The sign at the Sizzler was unlighted, as were all but traffic lights in the town. Inside the dimly lit restaurant we kept our coats on while we dined. It was the height of the 1978 energy shortage and we were attending a Home Appliance Technology

Conference at Purdue University. One of the subjects we discussed that night was the heat-pump water heater. Dennis was a researcher at Oak Ridge National Laboratory, and later that year would be the co-author of a paper that set off a controversy that still continues. The paper reported on a Department of Energy-sponsored study of water-heating options and concluded: "Heat-pump water heaters are likely to offer much larger energy and economic benefits than will solar systems, even with tax credits."

Studies by other agencies later came to similar conclusions. In response, the Solar Energy Research Institute (SERI) in Golden, Colorado, did a computer-model study that factored in the impact a heat-pump water heater has on space-heating loads to arrive at a "relative performance ratio" (RPR) rather than a COP. The SERI study made these assumptions: That the heat-pump water heater would use heat from the house rather than waste heat; that it would cost $1,000 to buy and install; that it would operate at an RPR of 1.6; and that it would need to be replaced in the tenth year.

To be more cost-effective, the study concluded, a solar water heater would have to cost less than $3,800 and supply more than 45 percent of the energy for water heating. And further, the "real discount rate" (the interest rate after inflation) could not be greater than three percent.

"If a 10 percent real discount rate is assumed," the report continues, "the solar water-heating system must cost less than $2,700 and supply more than 60 percent of the hot water."

What does it all mean? If you have the capital, ideal conditions (lots of sunshine, no long cloudy periods), and southern orientation, a high-efficiency solar heater may be the best long-term investment. But most of us don't meet those requirements, and the heat-pump water heater may

allow us to realize equal or better savings.

While solar water heaters qualify for federal tax credits, heat-pump water heaters do not, though the DOE has petitioned to have them included. Some local incentive programs do apply, however. Your electric utility can advise you.

## WHAT TYPE TO CHOOSE?

If you have a well that supplies water at 50 degrees F or warmer year-round and can provide a flow rate of two gallons per minute, you may want to do as Howard Patterson did and install a water-source heat-pump water heater. At present, only E-Tech makes one, and that is in limited production. It must be attached to an existing water heater, which serves as a storage tank and backup heat source. Also consider that the water must be reinjected or channeled to a drainage area. Reinjection (into a second well) is required in some areas.

If a water-source heat pump is not practical for you, you can get either the self-contained-type air-source unit or one that hooks up to your present water heater. This seems to offer a clear-cut choice: the add-on unit if your existing heater is in good condition, or the self-contained type for replacement or new construction. But it's not that simple. The add-on unit is less expensive (without tank) and is more flexible, since it can be located remotely from the storage tank. But you must have a standard water heater to serve as a storage tank. If you don't, you could buy a storage tank—but it may cost as much as a conventional water heater and it can't provide backup heat.

Self-contained models incorporate resistance elements to provide backup heat. That type also has the advantage of having no external wiring or piping: The heat exchanger is located within the tank,

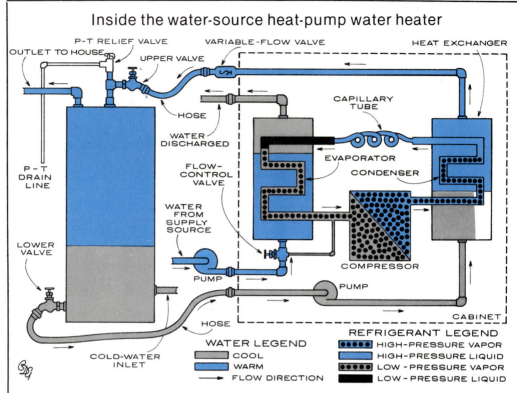

## Inside the water-source heat-pump water heater

E-Tech's water-source heat pump uses three separate circuits—a non-reversing refrigerant circuit and two water-flow circuits—to remove heat from supply water and transfer it to household water, which is held in storage.

An external pump forces the supply water through one tube of a double-tube heat exchanger (which acts as an evaporator), then discharges it. A liquid refrigerant at low pressure flows through the other tube. Heat from the water is absorbed by the refrigerant, vaporizing it. The vapor goes to the compressor, where its temperature and pressure are raised.

The hot, high-pressure vapor enters another double-tube heat exchanger, the condenser. A small internal pump circulates household water from the bottom of the storage tank through the other tube of the condenser and returns it to the top of the tank. The refrigerant vapor gives up heat to the water and thus condenses. The liquid flows into a capillary tube (a metering device), then back to the evaporator—and the cycle repeats. A thermostat actuates the system when water temperature in the storage tank falls below the setting.

so there's no heat loss from connecting pipes. Self-contained models also take up less total floor space. But they're more expensive and difficult to install.

Like any water-containing equipment, a heat-pump water heater must be protected from freezing temperatures. Most manufacturers recommend that their air-

## Air-source heat pump

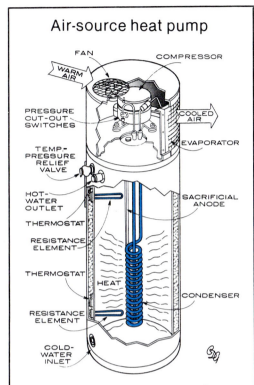

An air-source heat-pump water heater functions like the water-source system, but the refrigerant in the evaporator is warmed by room air, which is drawn over it by a fan. Shown is EUS's Temcor integrated water heater. Compressor and evaporator are in the upper compartment; condenser is in the water tank. Electric-resistance elements provide backup heat. Also see add-on version on page 310.

source units be installed where temperatures range from 45 to 95 degrees F. That eliminates outdoor and attic locations in most areas.

All models have external controls to protect the system from extremely high temperatures, and some have relays to transfer automatically to resistance heating if the temperature drops below 45 degrees F. The Oregon 12B units (see table) recirculate water from the conventional heater if the temperature drops to 34 degrees F to provide freeze protection.

Remember that an air-source heat pump absorbs heat from the area where it is located. The cooling and dehumidifying are a benefit in warm months but could be a liability when the heating system is in use. However, a heat-pump water heater won't dry out the air in your home when the relative humidity is below 60 percent, as it usually is in winter, since it operates well above the dew point.

You can reduce or eliminate the winter heat loss by proper location. Putting the heat pump in a basement or crawl space is usually best because at least some of the heat there comes from the ground. Placing it in a furnace room allows it to use waste heat.

A heat-pump water heater makes some noise, so you should not locate it near a sleeping area. For the same reason, I do not recommend suspending it from floor joists, as some manufacturers suggest, since vibrations could be amplified by the flooring.

Some provision must be made to connect the drain line that carries away condensate. If the unit is below the level of a drain, a condensate pump can be used.

A recent issue of a trade journal, *Air Conditioning & Refrigeration Business*, displayed this bold cover line: "Contractors: Get Ready to Replace 30 Million Electric Water Heaters." The editors are betting that a lot of people are ready to trade up to ↗ heat-pump water heater. They could well be right; you may never find a better investment.

# Buyer's guide to heat-pump water heaters

| Company | Model | Type* | Price ($) | Comments |
|---|---|---|---|---|
| Airtemp Corp.<br><br>Div. of Fedders Corp.<br>Edison NJ 08817 | AWH18A2B | A | 795 | Same specifications as Fedders. |
| Climatrol Sales Co.<br><br>Div. of Fedders Corp.<br>Edison NJ 08817 | CWH18A2B | A | 795 | Same specifications as Fedders. |
| Craftmaster<br>18450 S. Miles Rd.<br>Cleveland OH 44128 | n.a.<br>n.a.<br>HP 210C | I (66-gal.)<br>I (82-gal.)<br>A | 1,175 ⎫<br>1,275 ⎭<br>945 | Heat exchanger in top;<br>UL-listed; R-20 insulation.<br>UL-listed; thermostat in heat pump; delime mode. |
| Duo-Therm<br>509 S. Poplar St.<br>LaGrange IN 46761 | 41802-001 | A | 800 | UL-listed; delime mode |
| E-Tech<br>3570 American Dr.<br>Atlanta GA 30341 | W101<br>W102<br>101<br>102<br>B402<br><br>B403<br><br>B404 | A<br>A<br>A<br>A<br>A<br><br>A<br><br>A | 795<br>795<br>795<br>795<br>3,023<br><br>3,023<br><br>3,023 | Water-source; 3.5–4 COP.<br>220-V water-source; 3.5–4 COP.<br>UL-listed; DIY kit.<br>Same as above but 220-V.<br>3-hp commercial model; 220-V; 1-phase power supply<br>Commercial model; 220-V; 3-phase.<br>Commercial model; 440-V. |
| EUS (Temcor)<br><br>201 Seco Rd.<br>Monroeville PA 15146 | RE 825<br><br>RE 665<br>RE 505<br>RR115A<br><br><br>RR230A | I (82-gal.)<br><br>I (66-gal.)<br>I (55-gal.)<br>A<br><br><br>A | 1,100 ⎫<br><br>1,100 ⎬<br>1,100 ⎭<br>825<br><br><br>825 | Heat exchanger in tank, price includes installation; UL-listed.<br><br><br>Special plumbing connector lets water circulate from bottom.<br>Same as above but 220-V. |

# Buyer's guide to heat-pump water heaters (continued)

| Company | Model | Type* | Price ($) | Comments |
|---------|-------|-------|-----------|----------|
| Fedders Corp.<br><br>Edison NJ 08817 | FWH18A2B | A | 795 | DIY kit; thermostat in heat pump; UL-listed<br>140-deg., 130-deg., and 120-deg. settings |
| Mor-Flo Industries, Inc.<br><br>18450 S. Miles Rd.<br>Cleveland OH 44128 | HP 466<br><br>HP 482<br>HP 200C | I (66-gal.)<br><br>I (82-gal.)<br>A | 1,195 ⎫<br><br>1,295 ⎬<br>995 ⎭ | Similar to Craftmaster models but sold through contractors. |
| Northrup<br><br><br><br>302 Nichols Dr.<br>Hutchins TX 75141 | DHW-HP-10A | A | 650 | Dual blowers give quiet operation; thermostat in heat pump. |
| The Oregon Water Heater<br><br>8190 S.W. Nimbus Ave.<br>Building 5<br><br>Progress OR 97005 | 110-12A<br><br>220-12A<br>110-12B<br><br><br>220-12B<br>220-24A | A ⎫<br><br>A ⎪<br>A ⎬<br><br><br>A ⎪<br>A ⎭ | 1,795 (installed)<br>1,495 (w/o installation)<br><br><br><br>3,495 | Install where temperature is above 45 deg.<br>Same as above but 220-V.<br>Transfers to resistance below 45 deg.; gives freeze protection by recirculating water.<br>Same as above but 220-V.<br>High capacity; 220-V; install where temperature is above 45 deg. |

*A: add-on; I: integral. **Note:** All models are 110-volt unless otherwise specified. All units have 1,100- to 1,600-watt output with a COP of two or three unless otherwise noted. No attempt is made to compare them on this basis because there is no standardized test procedure; at this writing, a test standard, ARI 245, is being drafted.

# New, energy-saving heat-pump water heater

*Case history shows how to cut heating costs by more than half*

## By Evan Powell

That stippled-aluminum box you see in the photos may be one of the most significant developments yet in home energy conservation. Inside its unglamorous shell is a small heat pump designed to heat water for household use.

The Efficiency II heat-pump water heater, made by E-Tech, Inc., of Atlanta, heats water at the rate of 13,000 Btu per hour, equivalent to the output of a 3,800-watt high-recovery resistance heating element. But it consumes only 1,230 watts of electricity—less than one-third as much as a resistance element would use.

Heat-pump water heaters are not new; in fact, they appeared as early as the 1950's. But a powerful coalition of bargain-basement electric rates and shaky technology soon conspired to drive them from the market. Now, with electricity at luxury prices and heat-pump technology well-established, the concept has been revived. In 1978, a promising new integral heat-pump water heater was developed by Energy Utilization Systems under a subcontract with Oak Ridge National Laboratory for the Department of Energy. The E-Tech Efficiency II was the first to be commercially available. I installed one of the first, used it, and compared its energy consumption with that of a conventional electric heater (see box, "Keeping Score on Power Usage" on page 313). I was impressed with the savings.

Under many conditions, a heat pump will use less electricity to provide a given amount of heat—whether for space or water heating—than will an electric-resistance heater because it uses standard refrigeration techniques to absorb from the surroundings some of the heat needed to do the job (see box, "How the Heat-Pump Water Heater Works"). An electric-resistance element converts electricity to heat on a one-to-one basis: For each Btu of electric energy used you get out one Btu of heat energy. Engineers call this ratio of heat output to energy input the coefficient of performance (COP). With resistance heat, the COP is always 1.0. But a heat pump can have a higher COP because it uses electricity not to create heat, but to augment the heat it absorbs and pump that heat to another location. Under ideal conditions—68° F air temperature and 70° F inlet water temperature—E-Tech's heat-pump water heater will operate with a COP of 3.1. As

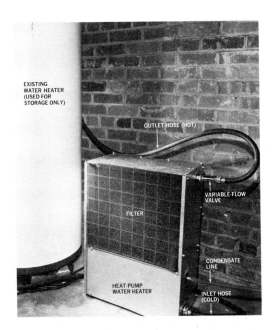

**E-Tech heat-pump water heater** connects to conventional electric water heater, which serves as a storage tank only. Thermostat cycles unit on and off to regulate tank temperature.

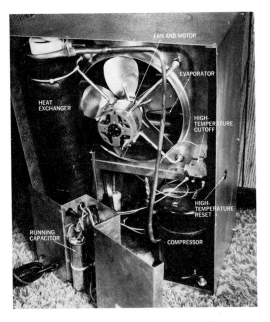

**Components are easily reached** for inspection or service with rear panel removed. High-temperature cutoff prevents system damage if water flow stops. Pump must be reset after cutoff.

air and inlet water temperatures go down, so does the COP. But the potential for saving is high, particularly since heating water uses more energy than any other household job except space heating.

## SOLAR SAVINGS—WITHOUT THE COST

Most experts agree that water heating is the most practical application for solar energy, but even a basic solar water heater is expensive and difficult to install. Most good solar systems could not surpass the performance I've realized with the Efficiency II heat-pump water heater, which can be bought at a fraction of the cost and installed in just a few hours.

A study sponsored by the Department of Energy and conducted by Oak Ridge

National Laboratory reached similar conclusions. The abstract of the report states: "Model results suggest that heat pump water heaters are likely to offer much larger energy and economic benefits than will solar systems, even with tax credits. This is because heat pumps provide about the same saving in electricity (about half) at a much lower capital cost ($700–$2,000) than do solar systems."

E-Tech, like Oak Ridge, estimates overall savings to be around 50 percent. My preliminary tests indicate that figure is conservative. Furthermore, the retail price is *lower* than the projections used in the Oak Ridge evaluation: The E-Tech production unit is $795 FOB Atlanta. A typical family of four may realize a saving of as much as $150 a year compared with a conventional electric water heater,

## How the energy-saving heat-pump water heater works

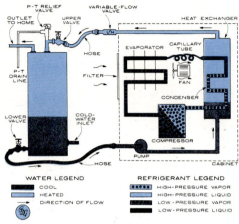

WATER LEGEND

| | |
|---|---|
| ▬▬ | COOL |
| ▬▬ | HEATED |
| → | DIRECTION OF FLOW |

REFRIGERANT LEGEND

| | |
|---|---|
| ●●●● | HIGH-PRESSURE VAPOR |
| ▬▬ | HIGH-PRESSURE LIQUID |
| ▬▬ | LOW-PRESSURE VAPOR |
| ▬▬ | LOW-PRESSURE LIQUID |

Two separate plumbing circuits—a nonreversing refrigerant circuit and a water circuit—are used by the Efficiency II to absorb heat from the ambient air and release it into the water. In the refrigerant circuit, a fan draws room air across the evaporator coil. The interior of the evaporator coil is a low-pressure area, and the heat in the air causes the refrigerant (a fluid selected for its low boiling point) to vaporize, absorbing heat.

The heat-laden vapor then goes to the compressor, where its temperature is further increased. The hot refrigerant vapor enters the condenser coil under high pressure. As heat is given off to the cooler surrounding water in the heat exchanger, the heat loss causes the refrigerant vapor to condense into a liquid. The liquid flows through the capillary tube, a tiny orifice that acts as a metering device, and back into the evaporator coil, where the cycle repeats itself.

In the water-flow circuit, a small pump circulates water from the bottom of the storage tank, through the heat exchanger in the Efficiency II, and returns it to the top of the storage tank.

The operation of the unit is controlled by a thermostat, which causes the compressor and pump to run when temperatures in the storage tank fall below the stat setting.

A variable-flow valve regulates the water flow rate to keep output temperatures at around 120° F. The positive-displacement recirculating pump tends to keep temperatures relatively consistent from top to bottom in the storage tank; a variation of 30–40° can occur with conventional water heaters.

---

indicating a payback period of less than four years.

At this point, the Efficiency II may sound like a partial solution to everyone's high electric bill—and it may be. But there's more to the story, and you should weigh the requirements and characteristics of the device against its merits before deciding if it's for you.

## DEMANDS OF EFFICIENCY II

First, consider the space requirements. The box measures 14 by 18 by 26 inches and weighs about 85 pounds. The do-it-yourself installation kit pictured here

allows a distance of several feet from the existing water heater, but the unit can be "hard-plumbed" with insulated pipes for permanent connection or for locating it more remotely.

Unlike space-heating heat pumps, with an outdoor evaporator coil and an indoor condenser coil, the Efficiency II is a one-piece device that must be located inside, where the temperature never falls below 45° F. At colder temperatures, frost would form on the evaporator coil. In very mild climates, it could be installed in an unheated space, but not in most parts of the country.

Since the unit works by removing some heat from the surrounding air and transferring it to the water, you should ask yourself where this heat comes from during winter months. If it's from a gas or oil furnace, you're okay in terms of cost per Btu. If that heat is coming from electric resistance units, however, you'd be gaining nothing by using the heat pump. On the other hand, if you're making use of waste heat (from a dryer vent, for instance, or from furnace-jacket losses), or if you locate the unit in a room warmed by a wood stove fueled with free firewood, you may reduce your water-heating costs by a significant amount.

In summer months, the cooling and de-humidification effect of the heat-pump water heater are beneficial. I noticed that my entire basement was free of damp and musty odors during the weeks the heat pump was in operation. But a condensate drain line must be provided to dispose of the moisture the heat pump removes from the air. If the unit is installed at a level below the house drains, a condensate pump will be required. Also, the Efficiency II's filter should be cleaned about once a month, depending on location.

The machine is not silent; there's a one-horsepower compressor and a recirculating pump inside that box, and it sounds very much like a window air conditioner. Located on a basement floor or in a crawl space, this should present no problem. But E-Tech also suggests hanging it from the ceiling or floor joists; there, harmonic vibrations may be propagated through the flooring. I would suggest avoiding this installation unless you are a heavy sleeper.

## INSTALLING IT YOURSELF

The only tools you'll need are a few screwdrivers, pliers, and a couple of pipe wrenches. The installation kit is complete even to the Teflon tape for the pipe threads.

The first step is to shut off power to your existing water heater, then drain the tank by attaching a hose to the boiler drain at the bottom and opening hot-water faucets in the house. As soon as the tank starts draining, you can install the top fitting. Remove the pressure-temperature relief valve and install the furnished T-fitting, nipple, and hand valve in the same opening. All modern pressure-temperature relief valves should have a handle that can be lifted for flushing. If yours doesn't, if the valve is more than five years old, or if the sensing element is too short, I would recommend replacing it with a new one.

Once the heater has drained, the old boiler drain at the bottom of the tank is removed and replaced by a long nipple and the new boiler drain.

The next step is to screw the variable-flow control valve into the unit's hot-water outlet and then connect the shorter of the two hoses provided between the bottom valve of the storage tank and the unit's cold-water inlet.

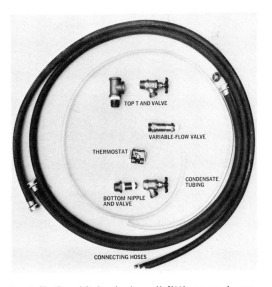

**Installation kit** includes all fittings and connections for normal installation. The pressure-temperature relief valve on existing tank must be replaced with a new one if sensing element is too short to extend into tank after adding a new T-fitting.

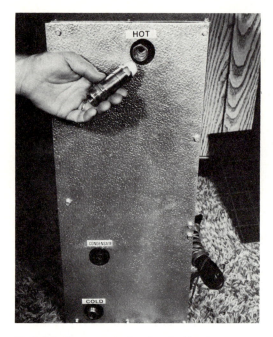

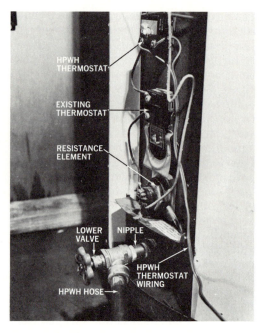

**Variable-flow control valve** in the hot-water outlet maintains 120° minimum water temperature by reducing the flow rate as necessary. Small amounts of hot water are available even if tank is cold.

**Lower boiler drain** on existing water heater is replaced with nipple and valve. New thermostat, installed above old one, has low-voltage circuit (24 VAC). Unit is available in 115- and 230-volt models.

Electrical connections are simple: Connect the supplied thermostat to the low-voltage wire that extends from the Efficiency II, remove the cover panel from the storage tank, and mount the new stat in good thermal contact with the tank just above the present bottom thermostat. The only deficiency I found with this entire system is the mounting of the new stat. E-Tech recommends using insulation to hold it in place, but I made a special bracket from an insulated wire loop, which I wedged under the two sides of the opening.

Next, clear plastic tubing is connected to the condensate outlet and run to a suitable drain or condensate pump.

Before connecting the hose from the hot outlet to the valve on the T-fitting at the relief valve, the unit should be purged of air by opening the bottom valve all the way and then plugging in the unit. After it runs for a few moments, the flow-control valve will open and allow the pump to force all air from the system. When the water runs clear, unplug the unit, connect the hose, then open both valves completely. Wait 10 minutes, plug in the heat-pump water heater, and the installation is done.

## SOLID GUARANTEE

E-Tech (3570 American Dr., Atlanta, Georgia 30341) offers a six-month, no-risk, money-back guarantee if you are not satisfied with the Efficiency II for any reason. And for five years the company will replace or repair any defective part. If you can't install the part yourself, it will repair, without charge, units returned to the factory, or an authorized dealer will make the necessary repair for only the cost of transportation.

The E-Tech unit I installed was one of the first production versions. The present models use an attractive enameled cabinet rather than the stippled aluminum on my first installation, and also incorporate a "sub-stat" temperature sensor with sensing tube that is inserted into the lower tank fitting rather than the tank-mounted thermostat.

At a time when boasts about the energy savings of many devices are greatly exaggerated, it's refreshing to find one that surpasses company claims. With a heat-pump water heater, you'll have no conspicuous array of solar collectors to announce to the world that you're saving energy, but you'll know you are when the electric bill arrives.

## Keeping score on power usage

**Kilowatt-hour meters** recorded the power consumption of the Efficiency II and a conventional electric-resistance heater. The heaters alternated weekly duty cycles. Here are the results:

| Week | HPWH (kWh) | Conventional (kWh) |
|------|------|------|
| 1 | 75 | — |
| 2 | — | 186 |
| 3 | 63 | — |
| 4 | — | 231 |
| 5 | 113 | — |
| 6 | — | 226 |
| 7 | 109 | — |
| 8 | — | 211 |
| 9 | 125 | — |
| 10 | — | 280 |
| Total | 485 | 1134 |

Duke Power Co. furnished and calibrated the two meters, and subsequently tested an Efficiency II in its Charlotte, N.C., laboratory. Initial testing indicates the unit has excellent recovery characteristics. Testing at 68° F ambient air temperature and 70° F water inlet temperature, hot water was drawn from a 50-gallon storage tank at the rate of three gallons per minute. At the factory thermostat setting, initial outlet temperature was 132° F. After eight gallons had been used it dropped to 130° F. After 50 gallons the water was still at 118° F.

# "Free" hot water from your air conditioner or heat pump

*Why buy more energy when you can use the heat your A/C throws away?*

## By V. Elaine Smay

All summer long your air conditioner gobbles up expensive energy as it pumps heat out of the house—while your water heater gobbles up more to heat your water. But wed those two appliances to each other and you can end this costly waste.

The box that can tie the knot is called a waste-heat water heater. Thousands of families, mostly in the South, have been using them for a number of years. Now, more and more companies are marketing them, including such major air-conditioner manufacturers as Carrier, Friedrich, and General Electric. The Department of Energy is eyeing them with interest, and utilities are promoting them as potential weapons in their battle to reduce peak summer loads. Water heating accounts for probably 15 to 20 percent of your household energy consumption. Perhaps you should consider arranging a marriage of your machines.

The heart of a waste-heat water heater is its heat exchanger, where hot refrigerant gas from the air conditioner's compressor gives up some of its superheat to cold water from the standard water

**Waste-heat water heater** (small box) made by ECU can be put outside; most must go inside for weather protection.

heater (see diagrams below and next page). Controls cut off the water (or refrigerant) flow when the temperature in the water tank reaches 140 to 160 degrees. As long as waste heat is adequate, the heating element in the regular water heater never comes on.

## HOW MUCH SAVINGS?

It depends on many factors, starting with the capacity of your A/C and how much it runs; you recover waste heat only when the compressor is on. Most manufacturers claim their systems can provide about 10 gallons of hot water per hour per ton of A/C capacity (but claims range from three to 18 gallons per ton-hour). At the 10 gal./T-hr. rate, a three-ton air conditioner could provide 30 gallons of hot water for every hour of compressor operation. If the compressor is on 25 percent of the time, for example, it would make 30 gallons of hot water in four hours. The table on page 316 gives kwh savings for seven cities, as calculated by General Electric. Dollar savings depends on how much you would have paid to heat the water in the standard way.

If the system is properly sized and installed, a waste-heat water heater can reduce the head pressure on the compressor; thus the A/C itself will use less energy—probably from six to eight percent less, according to most manufacturers.

## HEAT-PUMP HEAT RECOVERY

With a waste-heat water heater connected to a heat pump, you can also save on water-heating costs in winter. It's not "free" heat, but it's cheaper, at least if your water heater is electric. Here's why: A heat pump can deliver about twice as much heat for the energy it uses as can an

---

## How a waste-heat water heater works

**Typical waste-heat water heater** takes cold water from bottom of standard water heater and pumps it through a counterflow heat exchanger, where it picks up heat from hot (about 200 degrees F) refrigerant gas coming off the air-conditioner compressor. Refrigerant, cooled of some of its superheat, then goes to the condenser coil, where more heat is removed and gas condenses. Next, liquid refrigerant passes through expansion valve (not shown) and vaporizes in the evaporator coil inside the house, absorbing heat. Hot gas returns to the compressor and the cycle repeats. In the water loop, water passes from the heat exchanger into the top of the standard water heater. The cycle continues until water in the tank reaches a present maximum (140 to 160 degrees F). When the waste-heat water heater is hooked to a heat pump, it must be between the compressor and the reversing valve. With the heat pump in heating mode, the refrigerant cycle is reversed: The waste-heat device takes some heat that would have gone for space heating.

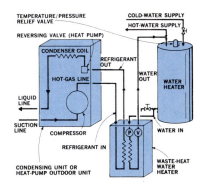

| City | Compressor on (hrs.) | Electricity saved (kWh) |
|------|------|------|
| Atlanta | 1146 | 2274 |
| Chicago | 782 | 1862 |
| Columbus, Ohio | 758 | 1777 |
| Dallas | 1595 | 2562 |
| Jacksonville, Fla. | 1675 | 2603 |
| Louisville | 998 | 2175 |
| Miami | 2448 | 2770 |

**Annual energy savings** for seven cities is based on using GE's Hot Water Bank with three-ton A/C. Numbers assume a family uses 75 gallons of hot water per day, that its temperature rise is from 70° to 140° and 30 percent losses occur in pipes and through jacket of water heater. In warmer climates, more waste heat is available than can be used, due to the 75 gallon ceiling on hot-water needs. Thus, savings per kwh are less than in cooler climates, though total savings are more.

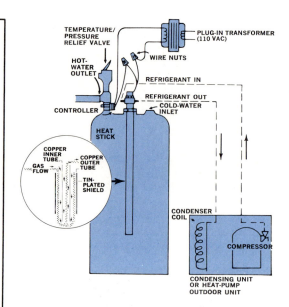

**Unusual waste-heat water heater,** the Heatstick, is installed through top of standard water heater by moving relief valve to hot-water outlet. Hot refrigerant gas from A/C enters via outer chamber of double-tube heat exchanger; exits up center.

electric-resistance heater (the kind in a standard water heater). Any time the heat pump's capacity exceeds the needs of the house (usually when outside temperature is above 30 to 35 degrees), the waste-heat water heater can take some heat from the heat pump, which in effect heats both house and water at the same high efficiency. The heat pump will just run a little longer. When the temperature goes lower, your regular water heater can take over.

Based on metered house installations, the Alabama Power Company concludes that these devices can provide approximately 50 percent of a family's hot-water needs when installed with a heat pump.

## HOW MUCH DO THEY COST?

Installed, about $400 to $800, depending on kind and installation. Arthur D. Little, Inc., did a study for the Electric Power Research Institute and concluded that the largest near-term market will be for heat-recovery units built into the A/C or heat pump, which should be cheaper.

"We arrived at $250 as kind of an average between the cost of an add-on and what we think it would cost if fully integrated," explained W. David Lee, who headed the study team.

Other considerations: "First I'd call my dealer and see if this thing would void the warranty on my air conditioner," says

Lee. Most manufacturers do honor the warranty as long as the heat-recovery unit is installed by trained personnel according to directions. Improper installation can adversely affect the A/C or heat pump.

Waste-heat water heaters are made in sizes for one- to five-ton residential air conditioners and heat pumps. (Larger commercial units are also available.) But selecting one is complicated by considerable fragmentation within the industry:

Manufacturers use different tests to calculate the energy saving, for example, and there's no agreement as to what kind of heat exchanger should be required to protect potable water from possible refrigerant leaks.

If these problems can be overcome, says the A. D. Little report, waste-heat water heaters could provide a saving equivalent to 42 million barrels of oil a year by 1990.

# Tankless water heaters—New way to lower your energy bill

*To take full advantage of these heaters, you may have to alter your habits*

## By Evan Powell

Heating water consumes more energy than any other job in the house except space heating. And no wonder: We use a lot of hot water and waste a lot of energy in the way we supply it. Most homes use a storage-type water heater—an insulated tank with an electric or gas heating element. It heats 40 to 80 gallons of water and keeps it hot day and night. And all the while, heat escapes through the tank's jacket.

Furthermore, most water heaters hold the water at a very high temperature—from 140 to 160 degrees F. Then it's mixed with cold water at the tap to cool it to about 98 degrees for most uses. "People would consider it ridiculous to set their space-heating thermostats to 90 degrees, then open doors and windows to cool the house to 70 degrees," says Robert Russell, president of Chronomite Laboratories. "But that's just what they do with hot water." And the higher the temperature of the stored water, the greater the potential for heat loss.

## THE TANKLESS ALTERNATIVE

One way to reduce hot-water energy waste is to use tankless water heaters. These heat the water instantaneously as it flows through the line. When you open the faucet, the heater comes on. When you close it, the heater shuts off. Since they don't store hot water, there are no standby heat losses.

An added benefit: Tankless heaters give you a constant supply of hot water. "Everyone in the family can take a shower—one after the other," points out Paul Grip, president of Tankless Heater Corporation, "and the last one will have as much hot water as the first." None will have the gusher he may be used to, however.

Tankless water heaters have been popular in Europe—where energy has long been costly—for years. Now they're coming to the U.S.

## HOW THEY WORK

The heating element in a tankless water heater is controlled by a flow-sensing device that turns the heat on at a particular minimum flow rate when you open a tap. Most operate with a constant heat input and a maximum flow rate, which depends partly on water pressure, and may be roughly adjustable by turning a knob on the unit. Flow rate and temperature rise are inversely related: The faster the flow rate, the cooler the water coming out. That's because it spends less time passing through the heat exchanger of the water heater, and, hence, absorbs less heat. True thermostatic models are to be introduced soon. They will regulate both heat input and water flow to deliver water at a predetermined temperature.

Tankless heaters are available in electric and natural-gas models (these can be converted easily to LP gas), and they can be used in two basic ways: as central units or in point-of-use installations. In a central installation they take the place of your existing water heater. This requires a large-capacity unit with a high flow rate and temperature rise (for the most part, only gas units meet these requirements), or several smaller units installed in series (gas and some electric units can be used).

In point-of-use installations, a tankless heater is installed near each appliance (or area) that requires hot water. Smaller-capacity heaters are usually sufficient.

Point-of-use heaters can also be used in combination with a storage-type heater. You can turn down the temperature of your storage tank to 110 degrees or so, for example, and use a point-of-use heater to boost the temperature at the dishwasher—the only appliance that needs 140 + -degree water—and at any outlets too far from the tank to maintain adequate temperature. Manufacturers call this a modified point-of-use installation.

**Tankless water heaters** save floor space. Gas models, such as this Thermar Homemaster, hang on the wall. Gas units need combustion air and must be vented.

Lowering overall hot-water temperature not only saves energy, it is also an excellent safety measure. Every year many people—mostly the elderly and the very young—are scalded by high-temperature household water.

There are also some specialized applications for tankless water heaters. The coming thermostatic models would be an ideal supplement for solar water-heating systems. Installed on the line between the solar storage tank and the house plumbing, they would heat the water or boost its temperature during long cloudy periods.

Small tankless heaters, such as the Thorn Nymph are excellent for campers. And the Stanton Flo Commander has a built-in timer that shuts it off—a fine feature for a child's lavatory for example.

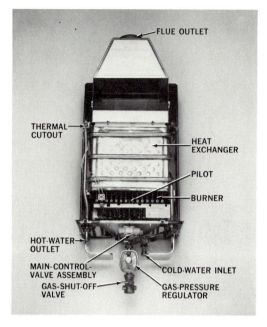

FLUE OUTLET

THERMAL
CUTOUT

HEAT
EXCHANGER

PILOT

BURNER

HOT-WATER
OUTLET

MAIN-CONTROL-
VALVE ASSEMBLY

COLD-WATER INLET

GAS-SHUT-OFF-
VALVE

GAS-PRESSURE
REGULATOR

**Gas tankless heaters** have a standing pilot that lights the burner when water flow is sufficient to close a switch. Direct-ignition models are planned.

## DOLLARS AND CENTS

It's impossible to put an exact dollars-and-cents figure on the savings you can expect by converting to instantaneous water heating. Every installation is different. And there has been no comparative study by an independent source. But some manufacturers' studies correlate closely with my own experience.

The Department of Energy sets the average family daily consumption at 64.3 gallons of hot water and the average temperature rise at 75 degrees. Temperature rise is the difference between the incoming-water temperature (which varies seasonally in most of the country) and the setting on the water heater. Using the DOE figures, Tankless Heater Corporation determined the cost of energy

required by standard storage-type 40-gallon electric, natural-gas, LP-gas, and oil-fired water heaters compared with tankless versions. The conclusion: Standing heat losses from storage-type heaters amounted to 32 cents per day for natural gas, 37 cents for electric, and 68 cents for LP gas. Furthermore, the tankless gas heaters were five percent more efficient in actually heating the water than the conventional gas heater, this study concluded. Thus, annual savings came to $125 with the natural-gas tankless unit (vs. the natural-gas storage heater), $135 for the tankless electric, and $267 for the tankless LP-gas heater.

Point-of-use heaters can potentially save more than central tankless units, because they eliminate standby losses from the water lines between tank and faucet as well as from the tank itself. Chronomite Laboratories did a study of a modified point-of-use installation, turning down the temperature of the storage heater from 160 to 110 degrees and using two of its Instant-Flow units as boosters. The study compared energy use with that of a conventional all-storage system. The result: The modified point-of-use system would save $78 a year compared with a conventional gas system, and $183 a year compared with an electric system.

"By simply adding small booster heaters at the 33 million dishwashers in homes in the U.S. and lowering thermostats on water heaters from 160 to 110 degrees F, America could save more than 92.3 billion cubic feet of natural gas, 24.9 million barrels of oil, and 27.8 billion pounds of coal per year," claims Chronomite's Russell.

In my own house, I've realized a 54 percent energy saving for water heating since I installed a point-of-use booster at the kitchen sink and dishwasher and reduced the storage-tank temperature to 120 degrees. At the same time, however, I installed water-saving shower heads and faucet flow controls, so I cannot say how much of the saving should be attributed simply to using *less* hot water.

The Tankless Heater Corporation study also points out the wisdom of using the fuel that gives the most heat per dollar. Natural gas is now the cheapest in terms of cost per Btu. (But it's not available everywhere.) The Tankless study shows that while its electric model, the Super Power Pack, could save $135 compared with a standard electric storage heater, it would cost $54 *more* per year to operate than a standard 40-gallon natural-gas heater.

## THE TRADE-OFFS

If you switch to tankless water heating, you'll probably need to adjust to lower water-flow rates. In the typical American home, the shower spews out from six to nine gallons of water per minute. Tankless heaters come in a wide range of capacities, but few can deliver an adequate temperature rise to give you a hot shower at that rate.

To get satisfactory performance, you would probably have to fit your showers with water-saving heads. These usually limit the flow to about 2.5 to three gpm. You might also want to put low-flow aerators (about one gpm) on your faucets. Even with these modifications, however, you might not be able to use more than two or three outlets at once drawing from the same heater.

Single-lever faucets can present a particular problem with tankless heaters. Since very little hot water flows through them on their warm setting it may not be enough to turn on the heater. If you have such faucets, you should choose a heater that has a low start-up flow rate.

Most large heaters have a control that lets you adjust the flow rate downward to provide a greater temperature rise in

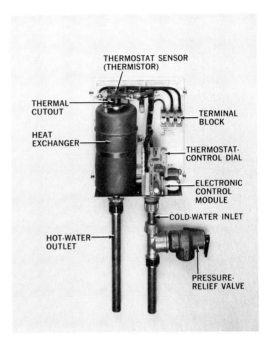

**Electric tankless heaters** are quite small and fit easily under a sink. With such a point-of-use installation, you get instant hot water—just turn on the tap.

winter. Measuring the temperature of your own water supply will give you a good idea of the temperature rise you need to provide 110-degree water, or whatever temperature you choose.

## MAKING THE SWITCH

The cost of a tankless heater could run from $180 for a small booster to $1,000 and up for a full conversion. Converting entirely to point-of-use heaters could be very expensive—each one costs about as much as a typical storage heater.

If your present water heater is in good condition, a modified point-of-use installation could save you money. (For more savings, you could wrap your storage tank in insulation.)

# Buyer's guide to tankless water heaters

| Manufacturer or distributor | Brand and model | Temperature rise (°F) @ flow rate (gpm)* | Max. heat input | Start-up flow rate (gpm) | Price ($) | Comments |
|---|---|---|---|---|---|---|
| **GAS** | | | | | | |
| Controlled Energy Corp. Box 19 Fiddler's Green Waitsfield VT 05673 | Saunier Duval 14 / 27 / 33 / 41 | 59 @ 1 / 40 @ 3 / 52 @ 3 / 63 @ 3 | 38,700 Btu/hr. / 77,400 Btu/hr. / 100,600 Btu/hr. / 123,000 Btu/hr. | 0.75 / 0.75 / 0.75 / 0.75 | 209.20 / 286.70 / 364.20 / 486.00 | SD-14 has optional tap, shower with flexible hose, and vent hood; SD-27 and -33 can take mixing taps; SD-41 has thermostat that regulates gas input according to incoming water temperature. All have a five-year warranty; American Gas Assn. approval pending |
| Environmental Research Assn. Box 531 Vineyard Haven MA 02568 | Thorn Nymph | 50 @ 1 | 30,000 Btu/hr. | 0.45 | 150 | Optional spout can be attached directly to heater. Good for campers and other special uses. Requires working pressure of only two psi. Piezoelectric igniters |
| ITS 7344-G S. Alton Way Englewood CO 80112 | Junkers 125 / W250 / W325 / W400 | 20 @ 3 / 43 @ 3 / 57 @ 3 / 67 @ 3 | 39,000 Btu/hr. / 75,000 Btu/hr. / 100,000 Btu/hr. / 117,000 Btu/hr. | 0.75 / 0.75 / 0.75 / 1.25 | 300 / 550 / 625 / 725 | Optional switch reduces start-up flow rate to 0.5 liter on models 125, W250, and W325, and to 0.4 liter on the W400. Junkers all have piezoelectric igniters |
| Paloma Ind. Inc. 241 James Street Bensonville IL 60106 | Paloma PH-6 / PH-12 / PH-16 / PH-24 | 24 @ 3 / 48 @ 3 / 65 @ 3 / 97 @ 3 | 43,800 Btu/hr. / 89,300 Btu/hr. / 121,500 Btu/hr. / 178,500 Btu/hr. | 1.7 / 1.4 / 1.9 / 2.9 | 290.52 / 549.60 / 796.18 / 893.73 | Flow regulator on heaters adjusts volume of water to regulate temperature. All have piezoelectric igniters. May not be suitable for use with single-lever faucets (see text) |
| Tankless Heater Corp. 20 Melrose Ave. Greenwich CT 06830 | Thorn Thermar Homemaster | 57 @ 3 | 103,400 Btu/hr. | 0.8 | 699 | External flow control can compensate for colder incoming water in winter. Modulating gas valve regulates gas input according to water-flow rate |
| **ELECTRIC** | | | | | | |
| Chronomite Laboratories 21011 S. Figueroa Carson CA 90745 | Instant-Flow MF 350 / MF 515 / MF 605 / S-23L / S-30L | 21 @ 3 / 31 @ 3 / 42 @ 3 / 16 @ 1 / 20 @ 1 | 9.2 kw @ 230 V / 13.8 kw @ 230 V / 18.4 kw @ 230 V / 2.3 kw @ 115 V / 3.0 kw @ 110 V | 0.75 / 0.75 / 0.75 / 0.3 / 0.3 | 180 / 180 / 180 / 180 / 180 | MF (Maxi-Flow) series is appropriate for central installations. These require special wiring. S-series models are small under-sink units; excellent as temperature boosters |
| ITS (address above) | ITS MDT-1.5 / MDT-3 / MDT-6 / MDT-9 | 14 @ 1 / 28 @ 1 / 44 @ 1 / 61 @ 1 | 1.5 kw @ 120 V / 3 kw @ 120 V / 6 kw @ 240 V / 9 kw @ 240 V | 0.25 / 0.25 / 0.25 / 0.25 | 160.50 / 160.50 / 160.50 / 170 | Small (cigar-box size) heaters that fit under sink in any position. Good as temperature booster |
| Tankless Heater Corp. (address above) | Stanton Super Power Pack / Elf / Heat Commander / Flo Commander | 68 @ 1 / 68 @ 1 / 68 @ 1 / 68 @ 1 | 7 kw @ 230 V / 7 kw @ 230 V / 7 kw @ 230 V / 7 kw @ 230 V | 0.5 / 0.5 / 0.5 / 0.5 | 220 / 220 / 220 / 220 | Flo Commander (also available as 4-kw model) has optional timer that shuts off water after specified interval. Built-in spray head serves as faucet or shower. Elf and Heat Commander have flexible hose and shower heads. Super Power Packs can be installed in series for central installations |
| Tri-American Products Corp. 1874 S.W. 16th Terrace Miami FL 33145 | Corona CTH 25 / CTH 30 / CTH 35 / CTH 40 | 25 @ 1 / 34 @ 1 / 45 @ 1 / 58 @ 1 | 3.6 kw @ 115 V / 4.8 kw @ 230 V / 6.5 kw @ 230 V / 8.4 kw @ 230 V | 0.35 / 0.35 / 0.35 / 0.35 | 180 / 180 / 180 / 180 | Small units that can be installed under a sink. They come with cord and plug for 110- or 220-V operation. Heat-exchanger cover comes off for cleaning. Flow-rate control on heater can be adjusted to give desired temperature |

*Temperature rise for heaters large enough for central installations is calculated at a three-gallon-per-minute flow rate. For smaller units intended for point-of-use installation, a one-gallon-per-minute flow rate is used. Some smaller units can be installed in series to serve as central heaters.

If you are willing to adjust your usage habits, a tankless gas heater might be a good choice when your old heater needs replacing. If your water line runs are short, consider a central heater. If they're long, point-of-use heaters might be worth the cost.

Before buying any replacement heater you'll want to consider other energy-saving alternatives, such as the newer super-insulated, high-efficiency storage-type units (I have the A.O. Smith Conservationist) and the new heat-pump water heaters. Where electricity is your best (or only) fuel choice, the heat-pump water heater may be the most economical alternative.

# High-efficiency gas water heater

## By V. Elaine Smay

In a conventional gas water heater, the gas burns at the bottom of the tank and hot combustion products travel up through a baffled vent in the center of the tank, heating the surrounding water. Of course, not all of the heat gets transferred to the water; much of it goes out the flue. And even when the burner isn't on, that vent bleeds heat from the surrounding water.

A new gas-water-heater design improves efficiency by putting the vent stack beside the tank rather than inside it. Water is heated as it circulates through a copper heat exchanger above the burner. The exhaust gases pass out the adjacent vent. The prototype was developed by Advanced Mechanical Technology, Inc. and Amtrol, Inc. under a Department of Energy contract supervised by Oak Ridge National Laboratory.

Tests of the unit (with a temperature rise of 90°F and an outlet temperature of 150°) show 66 percent efficiency, compared with 51 percent for conventional gas heaters and 61 percent for high-efficiency models.

Initial price would be from $115 to $175 above conventional models, but lower gas consumption would recoup the outlay in three to six years. The design should also outlast conventional water heaters, whose glass-lined tanks usually leak in 10 or 11 years. This one has a polyethylene liner and should last 15 years. Fifteen prototypes will undergo further tests. If they prove successful, the next step is commercialization.

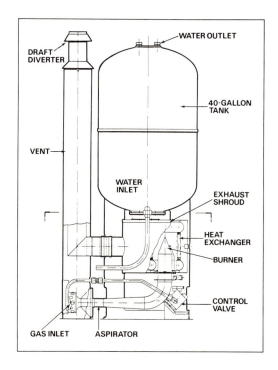

# 10 PORTABLE HEATERS

What about those portable heaters you see in all the stores these days? Do they really cost less to operate?

The answer is yes, but not necessarily directly in terms of more Btu's per dollar. This chapter will tell you how they should be used in a good program of zone or spot heating, allow-ing you to reduce your central heating system's running time by using them. It will also give you a good idea of the various types and brands that are avail-able, and the differences between them.

Thanks are due *Homeowners How To* for the round-up called "Heat Where and When You Want It."

# Portable radiant heaters—Do they cut heating costs?

---

*Quartz-electric and new kerosene radiants focus the heat for comfort at close ranges*

---

## By Evan Powell

Jim and Kay Miller live in a big old house that they've remodeled and weatherproofed. But their heating costs had gone through the ceiling anyway. "I solved part of the problem by installing a wood heater upstairs," says Jim. "But when we came downstairs we had to choose between freezing or feeding the thirsty oil burner. Now we're comfortable downstairs and we've cut our fuel bill in half."

Jim Hardwick lives in a ranch-style brick house. "My fuel-oil bill often ran $600 a year," he says. "But last year it was only $136."

Both homeowners give most of the credit for these savings to portable radiant heaters. They turned down the thermostats of their central-heating systems—to 50 or 55 degrees—and used the heaters to keep comfortable. The Millers'

solution was a radiant kerosene heater that Jim originally purchased for his greenhouse. Hardwick bought a quartz-electric heater.

Chances are you've been seeing quartz-electric and kerosene heaters in stores and ads. If you're considering buying one, you probably have a lot of questions. Do quartz-electric heaters really use less energy than standard electrics? Are kerosene heaters safe—and legal? Will a radiant space heater really keep you comfortable in a chilly room? Finally, will it reduce your heating costs?

Before we consider these questions, take a look at just what radiant heat is. Unlike convection heat, which warms the target by first warming the air around it, radiant heat travels in the form of infrared rays. When it strikes an object, it warms it. Ultimately, the air, too, will be warmed as heated objects give off heat.

## THE NEW RADIANTS

The "quartz" in radiant electric heaters is simply a tube that encapsulates a standard nichrome heating element. It protects the element from damage and also helps concentrate the heat, raising it some 500 degrees above the operating

**Quartz-electric heaters** emit infrared rays that warm objects, but not air. Reflector behind the quartz-sheathed element beams the rays at about a 30-degree angle. This is Markel's Quartz-Glow; its wide base resists tip-overs.

periods, say two or three hours. For longer periods, that advantage will probably lessen and eventually disappear.

Modern kerosene heaters bear little resemblance to those of 30 years ago. Airflow could not be closely controlled, and the kerosene often burned inefficiently. Result: high emissions.

Today's heaters, mostly Japanese-made, are precision-designed to deliver just the right amount of air to the fire. They burn with 99.9 percent efficiency, and thus low emissions, although definite safe levels have still not been fully determined. A wick, usually fiberglass, burns with a very hot blue flame, which

temperature of a conventional open-type heating element to around 2,000 degrees. Most such heaters are equipped with controls that pulse the power—it's on for a few seconds then off for a few. The quartz tube retains some heat, which makes the cycling less noticeable.

Will a radiant electric heater use less energy than a standard electric? Maybe—for some applications. Each kilowatt-hour of energy consumed will net 3,413 Btu of heat, regardless of the form it takes. But a convection heater in a large room may take hours to warm the air enough for the thermostat to cycle the element. With a radiant heater, you feel a warming effect almost immediately, and the cycling can begin at once.

For that reason, quartz-electric heaters will probably consume less energy than convection types when used for short

**Modern kerosene heaters** are finished with high-gloss enamel in handsome colors. The radiants usually direct the infrared beam over a 45- to 60-degree angle. This one, The Director from KeroSun, swivels for easier focusing of the heat.

heats a wire mesh or glass globe. Those elements emit infrared radiation. Actually, less than half of the heat that a radiant kerosene heater gives off is in the infrared band.

## BUT ARE THEY SAFE?

Unvented open-flame space heaters can deplete the oxygen in the air and discharge carbon monoxide. Many deaths occur each year because of carbon monoxide poisoning caused by such heaters, especially those that burn bottled gas or gasoline. Consequently, some local codes ban all unvented open-flame heaters. Manufacturers of kerosene heaters are lobbying hard—and often successfully— to get their heaters approved as exceptions in areas with such codes. Their claim is that these modern kerosene heaters burn the fuel so completely that they emit little CO and are thus safe *if used with a little ventilation.*

How much ventilation? Dr. Donald S. Lavery, a consultant in analytical chemistry and product safety, concluded in testimony to the Connecticut General Assembly that the oxygen requirement of a typical 10,000-Btu/hr. kerosene heater would be met if a window or door to an adjacent room were open one inch. Bruce Pershke, chief engineer at Koehring (and an enthusiastic believer in kerosene heaters), offers more conservative guidelines. "If you live in a tight house, leaving open a door to another room might have about as much benefit as putting a larger plastic bag over your head," he says. His recommendations: Open a window in the room where the heater is used, and don't use an open-flame unvented heater in a room where someone is sleeping. Other safety tips: No open-flame heater should be used where sawdust or other flammable material is present or on boats (see also page 338).

Carbon monoxide production is relatively low in the new kerosene heaters, but increases if the combustion air is reduced. U.S. standards for kerosene heaters are under development.

Many of the kerosene heaters are listed by Underwriters Laboratories, and most others are now undergoing tests there. Most of the imported heaters also meet the stringent standards of the Japan Heating Appliances Inspection Association.

Approved kerosene heaters have an anti-tip switch, which extinguishes the flame in the event of a tip-over. Some also have an earthquake switch, which responds to vibrations.

The quartz-electric heaters raise few safety questions. They all have an anti-tip switch and guards that help maintain a safe distance between the hot tube and your furnishings and have no emmissions.

## COMFORT AND MAINTENANCE

Will you be comfortable in an otherwise too-cool room if you use a radiant heater? First, you should know that a radiant heater is effective only within a 20 to 25-foot area; it's most effective within three to 10 feet.

A radiant heater gives a pleasant warmth, but I find it leaves me cold on the backside. If you're sitting on upholstered furniture, this is less noticeable, since the furnishings provide some insulation. You might also use two heaters.

There's little maintenance involved with the quartz heaters: Replacing the element is a minor chore. Kerosene heaters require more attention. First, there's the fueling. Several have a special pump and siphon to transfer the kerosene. Take the unit outdoors for refueling; it's impossible to refuel without residual kerosene odor. The burner needs occasional

# A sampling of kerosene heaters

| Manufacturer or distributor | Model | Heat output (Btu/hr.) | Tank capacity (gal.) | Approx. price($) | Comments |
|---|---|---|---|---|---|
| ALH Inc., P.O. Box 100255, Nashville TN 37210 | Aladdin Young II | 5700–7700 | 0.75 | 158 | Automatic ignition and shut-off. Ideal for small areas |
| | Aladdin Happy II Model J480 | 8000–9440 | 0.75 | 178 | Automatic shut-off; dual-tank construction |
| | Corona STDK | 9200 | 0.96 | 199 | Lift-out fuel tank with gauge, battery-powered ignition, built-in level. Uses 0.065 gal. of kerosene per hour |
| | Corona 22DK | 22,600 | 1.82 | 259 | Highest heat output of all portable kerosene heaters. Has built-in level, battery-powered ignition |
| | Toyokuni 26 EGR | 9000 | 1.10 | 159 | Glass-shielded burner is claimed to intensify heat output. Built-in drip tray catches spills. |
| Koehring, P.O. Box 719, Bowling Green KY 42101 | Comfort Glow GR8 | 9300 | 1.10 | 135 | Removable fuel tank; battery-powered ignition |
| | Comfort Glow GR9 | 9300 | 1.10 | 189 | Removable tank, fuel-level indicator, battery-powered ignition. Catalytic deodorizer minimizes odor during start-up and shutdown |
| | Radiant 10 | 9600 | 1.92 | 230 | Most popular radiant kerosene heater in the country. Runs 24 hours on full tank |
| | The Director | 11,700 | 1.99 | 270 | Burner head pivots left or right to direct heat. Has battery-powered igniter |
| | Moonlighter | 9000 | 1.7 | 170 | Convection heater—has transparent mantle that allows light and heat transmission all around. Top can serve as emergency cooking surface |

# A sampling of quartz-electric heaters

| Manufacturer or distributor | Model | Watts | Heat output (Btu/hr.) | Approx. price($) | Comments |
|---|---|---|---|---|---|
| Basic Accessories Corp., 1027 Newark Ave., Elizabeth NJ 07208 | Solar Beam Deluxe Q1001 | 700/400 | 2389/4778 | 80 | Two elements with switch to select single or dual operation |
| | 101 Portable | 1500 | 5119 | 90 | Rear-mounted thermostat, front-mounted control |
| | 301 and 301S | 1200 | 4096 | 70 | Wall-mounted quartz heaters mount on surface (301S) or can be recessed (301). Available for 115- or 230-V circuits |
| Hamilton Beach Div., Scovill Mfg., Washington NC 27889 | 208 | 1500 | 5119 | 97 | Front-mounted control, built-in thermal protector, all-steel base |
| | Quartz-Glo | 1500 | 5119 | 97 | Controller cycles element in proportion to variable setting. Oversize metal base and standoff help maintain proper clearances |
| McGraw Edison Co., 1801 N. Stadium Blvd., Columbia MO 65201 | Edison Quartz | 750/1500 | 2560/5119 | 77 | Horizontal heater has built-in fan to provide convective heat |
| | Presto Quartz | 1500 | 5119 | 94 | Top-mounted control; wide-angle parabolic reflector; dual-quartz tubes |
| Patton Electric Co., Inc., P.O. Box 128, 15012 Edgarton Rd., New Haven IN 46774 | QE2 | 400/800 | 1365/2730 | 80 | Dual elements give 400- or 800-watt output. Horizontal heater has built-in humidifier (10-oz. water reservoir) |
| | Ultra Quartz People Heater | 1400 | 4778 | 80 | Front-mounted control, long-life ½-inch quartz element. Sold by mail order. Plug-in tubes can be changed in seconds |

cleaning. Replacement of the wick can be rather involved, and is recommended each season with normal use. Replacement runs $15 to $30.

## WHICH TO BUY?

The quartz-electric heaters are lighter in weight—usually five to 10 pounds—and easier to carry from room to room than kerosene heaters, which may weigh 20 to 40 pounds.

Before you decide on a kerosene heater you should be sure kerosene is available in your area. Refineries may sell anything from aviation fuel to number-one fuel oil as kerosene. These fuels tend to have a high sulfur content, which you'll smell if you burn them in a heater.

High-quality kerosene should be crystal-clear—a yellowish tint usually indicates adulteration. Koehring (and perhaps other manufacturers) will test a sample you send in and tell you if it is appropriate for a heater. You can also check your heater after you've burned a tank or two. If you discover heavy tar deposits, find another source of fuel for the heater.

Another consideration may be your family's sensitivity to the odor of kerosene. Many people smell nothing at all while using the heaters, but some may notice a faint odor, especially at start-up.

In general, kerosene heaters put out more heat than electrics; they also cost more (see tables). Operating costs may be somewhat less, however. Kerosene sells for $1.14 a gallon in my area, although factory-packaged top quality fuel can cost more than $3.00. A typical 8000- to 10,000-Btu/hr. heater will run from 13 to 17 hours on a gallon of fuel.

Will a radiant space heater reduce your total heating bills? If you use it only occasionally, you may never recover the money you spent for it. But if you use it as part of a complete program of conservation, you could recover the cost in the first year.

# Heat where and when you want it

## By Evan Powell

A portable heater doesn't save energy in the sense that it has more output than a comparable conventional heater—any electric resistance heating element, for example, gives you the same 3.45 Btu for each watt of power consumed. But if the question is whether a portable heater can help you save energy in your home, the answer, based on two years of experimentation, is "yes."

There is a qualification: The heater must be part of an over-all household energy-saving program. If you buy a heater and make no other adjustment, you accomplish no more than raising your energy consumption by the amount the heater uses. But the savings can be substantial if you use a heater in conjunction with reducing heat input from your furnace and heating only occupied areas. You might, for example, start by maintaining the whole-house temperature at 60 degrees, which is surprisingly comfortable under active conditions; when you slow down and begin to feel cool, just turn on a portable heater. Of course, if there is a chronic illness in the family, a heat-reduction program of this kind should be undertaken only with the doctor's approval.

**Toyokuni 26EGQ**, is 8600 Btu per hour radiant kerosene heater. Has fuel capacity of 1.1 gallons, burn rate of .26 quart per hour. Other features include electric ignitor, automatic extinguishing, double safety tank and glass wool wick. From GLO International Corp., Dayton, OH 45402, which also produces the Corona 11-DK, a convection-type heater rated at 17,600 Btu per hour, with .53 quart per hour burn rate.

Other precautions:

What else should you look for? The accompanying photographs demonstrate the wide range of features available. But

**DeLonghi 930B,** above, operates at 600, 900 or 1500 watts. It is a permanently-sealed, oil-filled radiator; diathermic oil acts as heat sink, allows unit to radiate heat even after thermostat has shut off element. From De-Longhi-America, 350 Fifth Ave., New York, New York 10001. Flo-aire convection heater, below, has no moving parts, utilizes natural upward flow of warm air. Model FL 12 shown has switch-selected 600- or 1200-watt output. From Patton Electric Co., Inc., P.O. Box 128, 15012 Edgarton Rd., New Haven, IN.

with any heater you choose, look for heavy-duty construction, a large protective grille, tip-proof base, and the Underwriter's Laboratory seal on the cord as well as the heater.

## KEROSENE HEATERS

Although kerosene heaters differ from electrics in almost every other respect, they come in the same two types—radiants and convectors. They don't require any outside power source or hook-up. They are unvented heaters that demand only refueling with high-grade kerosene every day or two, depending on frequency of use. Most have removable tanks to ease this operation and help reduce odors. They also have a spring-loaded trip that shuts off the flame if the heater is jarred or turned over.

Although almost all kerosene heaters are touted to be odorless, if you're sensitive to the odor of burning kerosene, you'll probably detect it on start-up. If that's objectionable, look into a unit with a catalytic reactor. In testing a Koehring model, we found it to be much better from this standpoint than those without the catalyst, even when the catalyst was cold. Once in operation, none of the heaters should emit an objectionable odor. Neither the electric nor the kerosene radiant heaters should be used to heat sleeping quarters at night—they're intended for areas where you're living and working. From that standpoint they do a good job.

Though you'll see both kerosene and electric heaters advertised for use in basements and workshops, we wouldn't recommend the kerosene type for this purpose because of the open flame, and it's recommended that you use the electric heater only with extreme care. You certainly *don't* want to use either in close proximity to flammable materials, or when there may be solvents or flammable vapors in the area. Electric heaters shouldn't be used within reach of plumbing fixtures or pipes in a bathroom.

## SPECIALTY HEATERS

All sorts of new special-purpose heaters are also coming onto the market—for example, built-in glass-panel radiants that give a bath or dressing room instant

warmth and provide attractive decoration as well as downsized "jet" heaters to warm you while you work in the garage or even outdoors on a cold day.

But the secret to conserving energy with comfort with a portable heater is to make your selection carefully and use it wisely. Do that and you'll be among a majority of owners who are finding portable heaters a comfortable alternative to high heating bills.

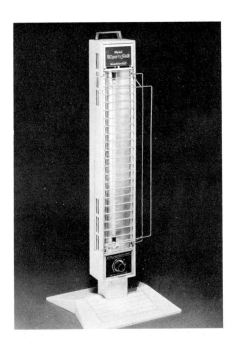

Quartz-Glo portable radiant unit, from Markel/NuTone (Madison & Red Bank Rds., Cincinnati, OH 45227), has wide-band reflector for broad heat distribution, tip switch, overheating device, 1500-watt single quartz tube. The Comfort Sensor, right, by McGraw Edison (1801 N. Stadium Blvd., Columbia, MO 65201) has 24-hour programmable timer, pilot light; choice of 1000-, 1500-watt output.

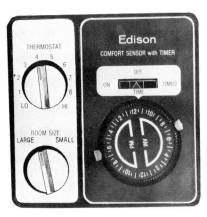

**Top-mounted switch** that controls dual elements at 600, 900 and 1500 watts; on/off indicator light; built-in handle and tip-over alert are among features of Deluxe model 4406, right, made by Northern Electric Company, division of Sunbeam, Chicago, IL 60625.

**Comfort Glow** model GQ15, above, by Koehring Atomaster Div., P.O. Box 719, Bowling Green, KY 42101, is 1500-watt, three-tube unit with built-in thermostat, safety switch and dual controls.

**Deluxe Model Q1001** by Basic Accessories Corp. (1027 Newark Ave., Elizabeth, N.J. 07208) is dual element heater rated at 700 watts on one tube, or 1400 watts on both.

**Among features** of Quartz Energy Saver Model 110 from Boekamp, Inc. (8221 Arjons Dr., San Diego, Cal. 92126): dual element switch that allows 750- or 1500-watt operation; cycle control; on/off light.

**Omni 15,** above, rated at 8700 Btu, holds 1.7 gallons of kerosene, has burn rate of .065 gallons per hour. Top surface can be used for cooking while camping or in power failures. Unit has an electric igniter and automatic tip-over switch. By Kero-Sun, Inc., Main Street, Kent, CI, 06757.

**Model SRP,** from TPI Corp. (P.O. Box T, Johnson City, Tenn. 37601) has heavy-duty construction, dual tubes that cycle together to provide 1500 watts of heat; cycling control for element pulsing; and automatic turnoff in the event of a tip-over.

**Vertical quartz heater** model 60H82, right, from Arvin Industries, Inc., Columbus, Ind. 47201, has top-mounted controls, two ½-inch diameter quartz tubes (output controlled at 750 or 1500 watts); on/off light; temperature limit, tip-over controls.

**Push-button switches** control heat at constant 600, 900 or 1500-watt levels in Quartzar model 1500 radiant heater by Klenatron (20 Hayward St., Ipswich, Mass. 01938). The handle and end caps are molded from Lexon.

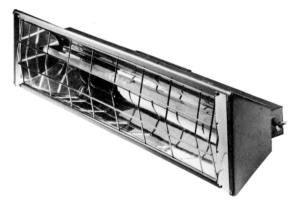

**Aladdin Temp-Rite 10** radiant heater model S481 has 11,300 Btu per hour output, holds .92 gallons of fuel. Has automatic lighting and shut-off and removable tank for easy refueling. Produced by ALH Inc., P.O. Box 10025, Nashville, Tenn. 37210.

**The Q27-2** quartz heater—from Martin Industries (P.O. Box 128, Florence, Ala. 35631)—is available with brackets for permanent mounting on a wall or ceiling. Such units are ideal for supplying radiant heat for patios, uninsulated areas or exposed and drafty spots around the home.

# Kerosene heaters—
# The safest type yet is one
# you must *install*

*Latest version of the controversial kerosene heater is vented to the outdoors for goof-proof operation*

## By Al Lees

A third generation of kerosene heaters should do much to dispel the clouded reputation that has clung to this widely used but often feared appliance. The unit being installed in the photo—a Monitor 30—answers most objections to old-style portables: It can't be tipped or moved too close to flammable fabrics; since it's vented, it won't burn up the room's oxygen supply; and it won't be improperly fueled once attached to its own outdoor tank.

Though more than 10 million households will buy kerosene heaters of various types through 1985, there are still homeowners who shudder at the very mention of them—and several states and cities still ban their use. Why? Because through improper use people still manage to kill themselves with kerosene heaters.

The fact is, according to the National Kerosene Heater Association (NKHA), there's no record of a single death or house fire caused by any modern kerosene heater, properly used. The bad reputation, NKHA insists, originally came from cheap, old-style units that were dangerous and smelly. It persists because of a few careless but tragic errors where users fueled one of the newer heaters with gasoline.

In these second-generation portables, pressureless wick-type systems draw kerosene from a tank placed as far away from the combustion chamber as the size of the heater permits. If the flame goes out, the fuel won't feed. (All heaters made or imported by NKHA members are UL listed.)

But new permanently installed vented heaters are intended as complete heating systems—ideal for room additions, converted garages, and attics. Last year, Kero-Sun (an innovator in this field that, ironically, declines membership in NKHA) introduced its Monitor 20, which preheats outdoor air for combustion as it exhausts byproducts (see diagram). It has push-button start-up, as well as a clock timer for automatic turn-on.

Now Kero-Sun has just introduced a bigger brother, labeled Monitor 30. It has

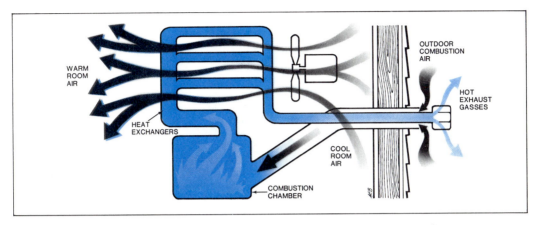

**Diagram shows** how vent system being installed, below, will work. Outside air is drawn in (around warm exhaust pipe) for combustion. Heat travels into series of heat exchangers; fan behind these draws in room air to blow through them.

**Through-the-wall installation** to vent heater (and connect it with outdoor fuel tank) is simple. We asked Kero-Sun for this demonstration with its new Monitor 30 to show that a 2½-in. hole is needed for the pipe-within-a-pipe, forced-flue-and-heat-exchanger system. Second, smaller hole will be drilled for fuel line to 55-gal. drum tucked inconspicuously into planter at right. Permanent installation eliminates main objections to kerosene portables—frequent refueling, possibility of oxygen depletion, and emission build-up.

a 13,000-Btu-greater heating capacity (32,600 compared with the 20's 19,600 rating) and offers new fuel-tank options that avoid the refilling nuisance: You can install a 55- or 275-gallon outdoor gravity-feed tank—or even a full-season underground tank. Timing mechanisms can automatically turn the heater on and off twice a day. Both Monitors claim a 92 percent fuel efficieny. (Venting does lose you a little: Most portables claim over 95 percent efficiency.)

Kerosene heating appliances are rated in size by the number of Btu they produce per hour—from 5,000 to over 30,000 Btu. To select the right size for a room, measure two walls to determine your square footage, then multiply that by 28 to get the rough Btu output needed. Prices range from $180 for a small portable to $989 for Kero-Sun's Monitor 30 (the Monitor 20 is $850).

Both vented heaters shown here and on the next page could be installed in an hour or two. Just drill a 2½-inch-diameter hole through an outside wall as shown. After mounting the vent-pipe assembly, drill a smaller hole for the fuel line and set the heater in place.

**Portable heater** can warm casual rooms such as this workshop. Kero-Sun's Moonlighter burns for 36 hours on a 1.7-gal. tank.

**Earlier, smaller version** of vented heater—Monitor 20—has a built-in capsule fuel tank that's removable for refilling.

## Portable kerosene heaters: Use with care

According to NKHA, imports on kerosene heaters climbed from 3,500 in 1974 to 3,154,000 in '81. The great majority of these new heaters are unvented portables that lack the features of the installed types. There are cautions you should consider before buying a portable.

The typical new portable is a compact unit with safety switches to shut it off if it's tipped or jarred (a carry-over from their Japanese origin, where the switch was required as an earthquake precaution) and a burner designed for safety and efficiency, using a wick that absorbs fuel by capillary action, located above a nonpressurized tank.

Most renewed interest in kerosene heaters has been spurred by energy conservation. The heaters are efficient: Almost all the fuel that's burned is converted into heat. But these units really come into their own when used for "zone" heating: You simply turn the central-heating thermostat down and use a kerosene heater to heat the room you are using. Savings can be spectacular.

Despite requirements for boldly printed decals on these heaters, many users ignore safety precautions such as proper ventilation. Due to the near-total combustion, emissions are low and odors are kept to a minimum, so it's easy to forget about oxygen depletion. Controversy surrounds the emission levels created by the heaters. Until the results are in, play it safe and ventilate. In a well-sealed room, these heaters can lower oxygen to a dangerous level. NKHA and some manufacturers suggest opening the door to an adjacent room. Others recommend (and I much prefer) cracking open an exterior window. If you locate the heater under the window, you'll get little draft.

*Never* fuel these heaters with anything but water-clear kerosene. No substitute will do—not aviation fuel, or No. 1 fuel oil, or "dirty" kerosene. It's especially important to guard against putting gasoline in the heaters. Store your fuel in a metal container clearly marked KEROSENE—and it shouldn't be a red can that could be filled with gasoline by mistake. Never refuel a heater indoors or when it's hot, and don't let children do the job.—*Evan Powell*

# NO
# MORE
# HOME
# POLLUTION

**P**ollution in the home? It can happen when you seal it up so tightly that the house can't "breathe," eliminating the substances that accumulate there. It's not a "no-win" battle, however, because you can control the ventilation in your home to prevent pollution without significant heat loss.

# Plugging all those heat leaks can cause home pollution

*Most houses are full of cracks that let in too much of the outdoors. But some may be too tight.*

By V. Elaine Smay

You've insulated to the hilt and aimed your caulking gun at every crack. You've weatherstripped around windows and doors—and even replaced some leaky double-hung windows with triple-glazed casement types. You've built vestibules at entries. No astronomical fuel bills this winter.

But reducing one problem may have worsened others. As we make houses tighter, more homes may be traps for moisture and worse—pollution.

"We don't really know how widespread the problem is," says Peter Cleary of Lawrence Berkeley Laboratory, where a number of Department of Energy-sponsored studies are going on. "Not enough houses have been monitored and no standards exist. Too often, health complaints are dismissed as psychosomatic."

But they may be real. Pollution levels in many homes are considerably above existing standards for outdoor air, some studies show. Tight houses are especially vulnerable, but even some with standard ventilation rates have been found to be polluted.

Solutions to the major problems are suggested in the information that follows. The problems include:

## Excess moisture in winter

In a leaky house, moist air from cooking, bathing, and human occupancy filters out and is replaced by cool, dry air that, when heated, has an increased capacity to hold moisture. Result: the relative humidity of the Sahara. But water vapor released in a tight house is trapped there. It can help break down building materials and reduce the effectiveness of insulation. Certain spores and bacteria thrive in high humidity, so illness may increase.

## Formaldehyde

A large number of products, including particleboard, chipboard, and urea-formaldehyde foam insulation, can emit formaldehyde gas. It is also a byproduct of combustion.

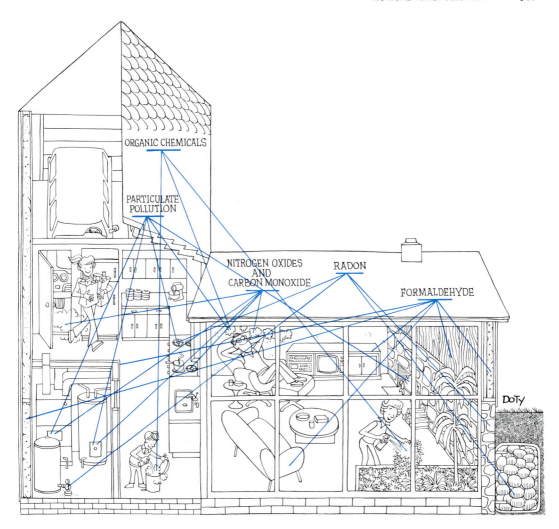

ORGANIC CHEMICALS

PARTICULATE POLLUTION

NITROGEN OXIDES AND CARBON MONOXIDE

RADON

FORMALDEHYDE

DOTY

Most people can detect the odor below one part per million. Concentrations of only half that can trigger swelling of mucous membranes. Higher levels can cause coughing, constriction of the chest, and a feeling of pressure in the head.

## Nitrogen oxides and carbon monoxide

In a kitchen-size test chamber at LBL, pollution sleuths ran the oven of a gas range at 350 degrees F for an hour and measured the emissions at various ventilation rates. Excessive concentrations of carbon monoxide occurred at low ventilation rates, but not when the fan was speeded up.

That same test oven produced excessive $NO_x$ levels at moderate ventilation rates. Field investigators have found concentrations above the recommended one-hour exposure standard for outdoor air in kitchens with gas ranges but without vented rangehood fans (or where they weren't used).

Carbon monoxide can cause headaches and dizziness at lower concentrations; nausea, vomiting, asphyxiation, and death at higher levels.

$NO_x$ increases the risk of respiratory problems. In England, a recent study found more such problems in children who lived in homes with gas ranges than in kids whose moms cooked electrically.

## Radon

Radon-222, a radioactive gas, is a decay product of radium-226, a trace element in most rocks and soil. It can enter the house through basements, slabs, and crawl spaces. Indoors, radon may emanate from concrete, brick, and other building materials. It is also found in ground water, natural gas, and LPG.

Radon's daughters (decay products) attach to particulates in the air and can enter your lungs. Many homes have measurable levels of radon. At low levels it probably causes no problems. But what happens in a tight house? No one knows for sure. And in areas with high levels of background radiation, radon concentrations may be high in a normally leaky house, too. Elevated levels of any radioactive substance are to be avoided, since at some undetermined level they can probably cause cancer.

## Other culprits

Combustion—tobacco smoking and burning wood, coal, and other fuels—produces particulates that can lodge in the lungs. Smoking a pack a day can cause concentration above the 24-hour standard for outdoor air.

Organic chemicals from smoking, pesticides, cleaning products, and hobby and crafts materials are potential pollutants. "We have a list of 35 or 40 organic chemicals found in indoor air," says Cleary. "Some are potentially harmful, some we know nothing about; and we certainly don't know their effects in combination with other chemicals."

Such a lack of solid information typifies the whole field of indoor air quality. But now, at least, we've begun to look at the problem.

# Blow out stale air but save the heat

---

*Heat exchangers you buy or build can do the job—in tight houses*

---

## By A. J. Hand

So you suspect your house of being polluted. Maybe you live in an area known to have high levels of radiation, or your house is so tight you have problems with excess humidity. Maybe you smell formaldehyde, or your family has a lot of colds. What should you do?

The solution that can deal most effectively with the widest range of indoor pollutants is to ventilate your house with a fan and get rid of the pollutants before they accumulate. But heating all that cold, fresh air can eat up a lot of fuel and money—exactly the kind of waste you sought to avoid by tightening your house.

But there is a device that lets you ventilate a tight house and save the heat: It's called an air-to-air heat exchanger. It transfers heat from your warm exhaust air to the cold air coming in. A fan feeds the air through a heat-exchange core—closely spaced, alternating intake and exhaust passages. Usually the fan draws stale, warm air out of the house through the exhaust passages. Meanwhile, cold outdoor air is drawn in through the intake passages, picking up heat from the air going out. The system is useful during the air-conditioning season, too: Incoming hot air is precooled by outgoing cool air.

## GETTING CHECKED OUT

Before you take any action, you'll want to confirm that your house is indeed polluted. Peter Cleary of Lawrence Berkeley Labs advises: "Contact the county health department in a large metropolitan area or the state public health department in less populated areas. If you think your problem may be radon, contact your local environmental protection agency. If you think the pollution comes from a consumer product, notify the U.S. Consumer Product Safety Commission. It's possible you'll get the run-around," Cleary cautions. "These agencies have had so few complaints so far that they may not know which agency is responsible for which complaints."

But with persistence, one of these agencies—possibly with the help of others—will inspect your home and either confirm or deny your suspicions.

(Commercially available gas-leak detectors, designed to detect the presence of propane and natural gas, can also alert you to the buildup of fumes from solvents

343

and paints. They may even detect excessive concentrations of carbon monoxide. They are not UL-approved for these purposes, however.)

If your fears of pollution are confirmed, you then face another problem: How much ventilation do you need to clear the air? "That can vary considerably from house to house," says Gary Roseme of LBL, "depending on the strength of the source of pollution." But field tests done by LBL may offer some reasonable guidance.

A standard measure of infiltration (or ventilation) rates is the air change per hour (ach). One ach means that a quantity of air equal to the total volume of your house leaks in—or is drawn in—each hour. "In houses where we have measured less than one-half ach, we have found pollutants," reports Roseme. "When we increased the ventilation rate to one ach, we have been able to reduce the concentration to acceptable levels."

## HOME HEAT EXCHANGERS

Ventilation systems with heat exchangers have been used in commercial buildings for years, but the idea of using them in homes is so new in this country that, as I write this, there are very few available of the size, price, and performance suitable for home use.

According to the University of Saskatchewan's R. W. Besant, a designer of heat exchangers, "You need a unit that costs under $500 installed or it won't pay."

According to my research, there are three possibilities below that price:

- A do-it-yourself design developed by Besant. It's also available ready-made from D. C. Heat Exchangers in Saskatoon, Sask., for around $425.
- A Japanese import, the $250 Lossnay, by Mitsubishi.

- The Z-Duct, being developed by Des Champs Labs in New Jersey.

Each of these heat exchangers has its problems, however. The Lossnay has very narrow air passages that can frost up and clog in cold weather. Besant has tested it and says, "It's not really built for a difficult environment." And Roseme, who is LBL's heat-exchanger specialist, has questions about the Lossnay's permeable-paper exchange surfaces: "In a very cold climate with a tight house and very high indoor moisture levels, you'd be bringing that moisture right back in. Also, there is the question of whether the water-soluble pollutants would come back."

Eugene Leger, designer of the "Double-Wall House" in northern Massachusetts, used the Lossnay last winter from December on, however, and reports that no frosting occurred. He also mentions a bonus feature of the Lossnay: "Since the core is moisture-permeable, latent heat as well as sensible heat can be recovered."

The Besant design? Besant himself admits that "the polyethylene we use as the heat-transfer surface tends to be a bit floppy." He's working on a new design using rigid plastic. The Besant has larger air passages than the Lossnay and can run even when frosted. And it can be defrosted with a solar air preheater (see drawing).

The Des Champs Z-Duct exchanger is now at the prototype stage. If and when it comes on the market, it could be the best of the three. It is made of aluminum and designed for cold climates.

## HOW WELL DO THEY WORK?

Aside from design problems, there's disagreement as to the effectiveness of any heat exchanger in a nontight house. There's no guarantee that the intake air will pass through the exchanger. Some of

# You can make your own heat exchanger

**Besant exchanger** is made of a 4 x 8 plywood sheet, polyethylene film, galvanized nails. Laminations separate layers of polyethylene (37 in all) by ½ inch. Variable-speed fan (150 cfm), ductwork, two-inch rigid fiberglass insulation are left out for clarity. Moisture will condense on polyethylene and drip into drain pan below. In cold weather, frost will form. To defrost, rotate three dampers to send exhaust air in other direction. Alternative: solar pre-heater shown on next page.

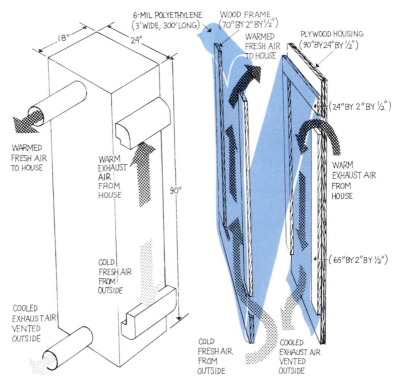

6-MIL POLYETHYLENE (3'WIDE, 300'LONG)
WOOD FRAME (70"BY 2"BY ½")
PLYWOOD HOUSING (90"BY 24"BY ½")
WARMED FRESH AIR TO HOUSE
18"
24"
(24"BY 2"BY ½")
WARMED FRESH AIR TO HOUSE
WARM EXHAUST AIR FROM HOUSE
90"
WARM EXHAUST AIR FROM HOUSE
(65"BY 2"BY ½")
COLD FRESH AIR FROM OUTSIDE
COOLED EXHAUST AIR VENTED OUTSIDE
COLD FRESH AIR FROM OUTSIDE
COOLED EXHAUST AIR VENTED OUTSIDE

it will, of course, but some won't. The rest will just enter as usual, around windows and doors, through cracks, and down the chimney. The looser the house, the lower the percentage of air passing through the heat exchanger.

According to Besant, "Most homes leak at such a rate that no heat exchanger is justifiable." Besant feels that any home being fitted with a heat exchanger should have vestibule entrances with two doors; sealed, casement, or awning windows (no doublehung windows); a complete vapor barrier in walls, floors, and ceilings; and provisions to isolate combustion air from the rest of the house.

How many homes in the U.S. could benefit from a heat exchanger? Fulton Cooke of Flakt Products, an importer of commercial heat exchangers from Sweden, says, "I would doubt if it were more than ¹⁄₁₀₀ of one percent."

"I don't agree," says Roseme. "Sure, there aren't a lot of very tight houses where these could be *best* applied, but there are a fairly large number—even though they aren't that tight—where pollution levels are high enough that running some air through would make sense, and using a heat exchanger could recover some of the heat."

All considered, mechanical ventilation seems to be the best approach to fighting indoor air pollution. Besant sums it up this way: "The answer lies in ventilating actively by fans. Whether one puts that

## A solar preheater for the DIY heat exchanger

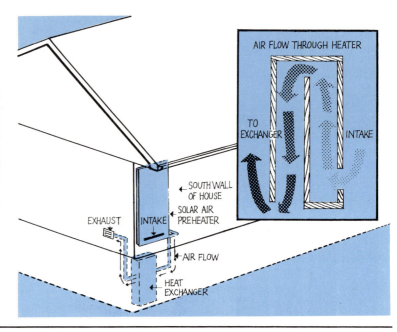

**To avoid frost buildup** in winter, the Besant heat exchanger can use a small solar collector to preheat the intake air. Plans for the collector are included when you order the heat-exchanger instructions.

AIR FLOW THROUGH HEATER

TO EXCHANGER

INTAKE

SOUTH WALL OF HOUSE

SOLAR AIR PREHEATER

EXHAUST

INTAKE

AIR FLOW

HEAT EXCHANGER

air through a heat exchanger or not depends on economics.'' (See box to calculate the economics for your home.)

## OTHER CONTROL METHODS

There may be alternative ways to avoid certain pollution problems and still save the heat. I talked with Dick Jewell, a Weyerhaeuser chemist who is working on the problem of formaldehyde emissions from plywood and particleboard. ''If your problem is particleboard,'' he says, ''a good shellac will help. Plywood paneling can be tougher, because if you put a sealer on it, you may change its appearance—especially with rough-sawn textures. Most producers are working on the formaldehyde problem, and you

should start seeing new products on the market pretty soon.'' These new products may be sealed, or they may be fumigated with ammonia in a process that chemically changes the formaldehyde. According to Jewell, they will almost certainly be labeled and clearly identifiable.

If your problem is humidity, it can be solved by venting at the source (in bathrooms, for example) by dehumidifiers, and with a careful eye to moisture-producing activities.

With gas ranges, the best solution is to use a range-hood fan whenever you cook. It will easily take care of carbon-monoxide emissions. Nitrogen oxides, however, are more difficult. According to measurements taken in a test kitchen at Lawrence Berkeley Labs, ''With a ventilation rate of 50 cubic feet per minute,

# Compute your own heat-exchanger economics

**How much money** would it cost you to heat all the cold air a power ventilator would blow through your home during a heating season? Here's a simple formula to help you find out: 0.025 x deg. day x cfm x fuel price = dollars per season. In this formula, *deg. day* is the degree-day rating for your geographical location, *cfm* is the proposed ventilation rate in cubic feet per minute (as fans are rated), and *fuel price* is your cost per 1000 Btu as given in the table.

Here's an example for a home in Boston (5634 deg. day) with a ventilation rate of 140 cfm, and a fuel price of a dollar a gallon: 0.025 x 5634 x 140 x .01 = $197.19.

This home would spend almost $200 heating fresh air for ventilation. Would a $500 heat exchanger be justified? If the house is fairly tight, about 75 percent of the intake air might pass through the exchanger. Figure the heat exchanger to have an efficiency of 75 percent. Multiply the cost per season ($197.19) times the amount of air passing through the exchanger (75 percent), times its efficiency (also 75 percent). The answer ($110.92) is the amount the exchanger could save.

|  | $ per kWh | $ per 1000 Btu |
|---|---|---|
| Electricity | .035 | .0103 |
|  | .04 | .0118 |
|  | .045 | .0132 |
|  | .05 | .0147 |
|  | .055 | .0162 |
|  | .06 | .0176 |
|  | .065 | .0191 |
|  | .07 | .0206 |
|  | $ per gallon | $ per 1000 Btu |
| Fuel Oil | .85 | .0085 |
|  | .90 | .0090 |
|  | .95 | .0095 |
|  | 1.00 | .0100 |
|  | 1.05 | .0105 |
|  | 1.10 | .110 |
|  | 1.15 | .0115 |
|  | 1.20 | .0120 |
|  | $ per 1000 cu. ft. | $ per 1000 Btu |
| Natural Gas | 2.25 | .0032 |
|  | 2.50 | .0036 |
|  | 2.75 | .0039 |
|  | 3.00 | .0043 |
|  | 3.25 | .0046 |
|  | 3.50 | .0050 |
|  | 3.75 | .0053 |
|  | 4.00 | .0057 |

the upper limit of recommended kitchen ventilation rates, the nitrogen dioxide concentration in the kitchen was considerably higher than promulgated standards."

However, the LBL kitchen is a small, enclosed room. Experiments in real kitchens have shown that using vented range-hood fans at high speeds (on the order of 100 cubic feet per minute) can control $NO_x$.

With a vented combustion appliance (gas or oil furnace, wood or coal stove, or a gas clothes dryer, for example), proper installation and regular professional servicing should keep them operating cleanly (But see "Oil Burner Hazards," Chapter 2, for potential problems even with regular servicing). In a tight house, combustion air from the outside may be necessary.

Radon is an even more insidious

| Heat exchanger vs. simple power ventilator—how much savings? | | | | | | |
| --- | --- | --- | --- | --- | --- | --- |
| | 0.5 air change per hour | | | 0.75 air change per hour | | |
| City | Oil | Gas | Electricity | Oil | Gas | Electricity |
| Atlanta, Ga. | $ 34 | $ 25 | $112 | $ 24 | $16 | $ 92 |
| Chicago, Ill. | 85 | 65 | 264 | 69 | 51 | 224 |
| Minneapolis, Minn. | 138 | 106 | 421 | 115 | 87 | 360 |
| Washington, D.C. | 52 | 39 | 166 | 40 | 29 | 139 |

**Lawrence Berkeley Labs** determined how much money a heat exchanger might save per heating season for a house in four cities. The LBL study found the heating cost for a house with 1800 sq. ft. of floor space, double-glazed windows, R-38 ceilings, R-19 walls, and a natural infiltration rate of 0.75 air change per hour (ach). This was then compared with the cost of heating the same house tightened to 0.2 ach, and vented through a 75 percent-efficient heat exchanger at two different rates. These rates, 0.5 and 0.75 ach, include the 0.2-ach natural infiltration. Gas and oil furnaces are assumed to be 70 percent efficient; electric, 100 percent.

problem. To begin with, if you suspect radon pollution, you'll be hard pressed to have it checked out. According to Bill Nazaroff, LBL's expert on instruments for radon detection, "At this point there are no readily available facilities for doing that." A further problem: A single measurement may be misleading. The concentration may vary greatly from room to room and from day to day.

Says Peter Cleary: "If you live in a very tight house in an area known to have high radium levels—near a uranium mining area, for example, or where the water is known to contain large quantities of radon—then you could assume your house would be more likely to have high levels of radon. But," he cautions, "while you can make that assumption, not enough large-scale surveys have been done to say it with certainty."

Though there are many potential sources of radon within the home (see preceding article), current evidence suggests that where there are elevated levels, radon has usually entered as a constituent of soil gas, seeping in through tiny cracks in the basement, through an unvented crawl space, or through the slab. With a crawl space or unheated basement, direct ventilation of that area can solve the problem. If the basement is heated, a heat exchanger may be justified. If so, it may as well serve the whole house.

Sealing the basement is another approach to radon control being tried in Canada. Arthur Scott of DSMA Atcon, an engineering firm with nuclear experience, has had some success sealing cracks and other entry points such as gaps around pipes and wiring with a urethane-rubber compound. Over that goes a protective layer of glass-reinforced concrete.

The future? Everyone agrees that as more and more houses get tighter and tighter the heat-exchanger market will open up. Right now there just isn't enough demand to encourage mass production. Nick Des Champs, president of Des Champs Labs, says that if he could sell 500 at a time, it would be possible to get the price down to $100 to $150. "Now that's getting in the ballpark of an ordinary exhaust fan."

Meanwhile, work goes on in other areas. Lawrence Berkeley Labs is studying the effectiveness of air-cleaning devices: electrostatic precipitators and high-efficiency filters, already in use in many homes to control particulates. LBL is also looking at air washers and filtering systems similar to those used in commerical buildings. If your outside air is also polluted, such devices may be your only route to healthy indoor air. But no one yet knows what pollutants they'll remove or how effectively they'll do it.

# Two routes to summer cooling

*Keep the air moving with a whole-house fan or wind-driven roof turbine*

## By Daniel Ruby and Louis Hochman

The lazy way to keep your home cool is to crank up the air conditioner. That's also the most expensive and least efficient way. A better approach is to intsall an inexpensive ventilation system that prevents the buildup of heat and humidity in the first place—so that you'll need air conditioning only when it's really steaming outside.

Many houses get uncomfortably hot because air in the attic is permitted to stagnate and collect heat as the sun beats down on the roof. Even if the attic is perfectly sealed, much of that heat is transfered by conduction and radiation to the living area below. (A related problem occurs in winter when unvented attics allow condensed moisture to accumulate.)

Solution: Air out your attic. Two ef- fective ways to do this are with a wind-driven roof turbine or a whole-house fan. A turbine moves air through the attic only, but it indirectly lowers temperatures throughout the house. Its vanes are powered by the wind, so it cools your house with no impact on your electric bill. Whole-house fans are electric-powered, but the watt-hours they use are insignificant next to the power that air conditioners consume. The fan mounts in your attic, but it draws in air through open windows in the living area, so it has a direct cooling effect. (Never operate a whole-house fan without opening a window—especially if you have gas appliances. Negative pressure can snuff pilot lights.)

Both systems require suitable vents for intake or exhuast. When installing either system, make sure that your attic has sufficient under-eave, gable, or ridge vents.

The picture sequences run through sample installations. Both can be homeowner jobs: The Emerson fan pictured requires no cutting of joists or framing. Emerson fans are made in three sizes—for houses ranging from 1,200 to 2,300 square feet. They are available at hardware and home-center stores for about $160 to $200.

Our roof ventilator is a 14-inch aluminum unit from Sears (approximately $40). Its lightweight construction makes its vanes sensitive to the slightest breezes— not the case for heavier galvanized models. Two turbines were installed by co-author Hochman to handle an 1,800-square foot house— positioned a quarter the length of the roof from either end and three feet below the peak of the roof so that the turbine tops are exposed to wind from all directions. (Lucky Installations of Van Nuys, California, offered some tips that made the job easier and the results neater.)

To find the center between rafters, make a saber-saw cut through the roof at the point where you want to mount the unit. Continue the cut horizontally until the blade contacts a rafter, then turn the saw around and cut the other way until the blade hits the opposite rafter.

After the turbine's base plate is installed over the hole, replace loosened shingles so that uncut pieces go under the plate and cut sections go over it. This ensures that rain will run off without leaks.

## 1. Install a roof turbine

**Wind-driven roof turbine** cools home indirectly by venting attic. Stale air goes out turbine; fresh comes in under-eave vents.

**Clipping the corners** off the turbine's base plate will ease the later job of sliding the unit under the loosened asphalt shingles.

**Find the center** between rafters as described in the text. Then position the base and trace around its inside circumference.

**To cut circle,** first saw small starter hole to serve as a handhold for removing cutout sections. Then cut two semicircles.

**Remove nails** from the shingles around the hole and slip base underneath until snugly in place. Replace shingles (see text).

**Drive large, galvanized** nails through the base plate on all sides to anchor the unit. A dab of tar on each nailhead prevents leaks.

**Align the unit** by spinning the neck sections and checking with a carpenter's level. Secure the turbine to the base with screws.

# 2. Install a whole-house fan

Whole-house fan cools home by drawing in fresh air through open windows while expelling hot, stale air through attic vents.

**To install Emerson fan,** decide where to locate it (centrally, if possible), then scribe ceiling cut with supplied template.

**Cut out scribed section;** lift fan through hole. Screw supports to joists. There's no need to cut joists or build special framing.

**Supply 120-V AC line power** per local codes, and wire to fan. Cut slots in plenum sleeve liner and attach with snaps.

**Screw lightweight,** louvered shutter in place over the ceiling opening. Shutter opens and closes automatically with fan use.

# Best route to summer cooling? Install a whole-house fan

*A ceiling mount does double work: It forces out hot attic air while drawing in fresh*

## By Bill Hawkins

Put an air conditioner in the window and it cools the room. Turn on a central air conditioner and it will cool your heels when the electric bill arrives. But install a whole-house fan and you've lowered the temperature of your entire home—at half the running cost of a window air conditioner.

The idea is simple: Remove the hot, stagnant air from the attic and, at the same time, draw in cool air throughout the house below. To do that, two types of fans are available. A horizontal mount is simplest and most effective. A vertical mount, however, can be used as a whole-house fan or, with its trap door shut, as an attic fan to work with your present air conditioners (see drawings, page 357).

How well do they work? During the hottest days, nothing cools like an air con-

ditioner. You'll get the most benefit out of a fan during the milder spring and fall months—when an air conditioner is simply overkill.

The fans come in three blade sizes: 24, 30, and 36 inches. Naturally, the larger the blade, the more pulling power it will have.

Fans are rated in cfm—cubic feet of air moved per minute. So, to choose the correct size, you must first calculate the number of cubic feet in your home, excluding closets, storage areas, and the attic. Generally, if you live in the South, a complete air exchange every minute would be desirable. In the cooler northern part of the country, an air exchange about every two minutes would be sufficient.

Shutters prevent air flow between the attic and living areas when the fan is off (especially during winter). They're automatic: Suction when the fan is on draws the shutters open, allowing air to flow. Switch the fan off, and they spring shut.

The type of shutter you use—double or single—depends upon your house construction. With a truss roof (diagonal framing between attic roof beams and attic flooring), you can't cut out any crossbeams without complications (like your

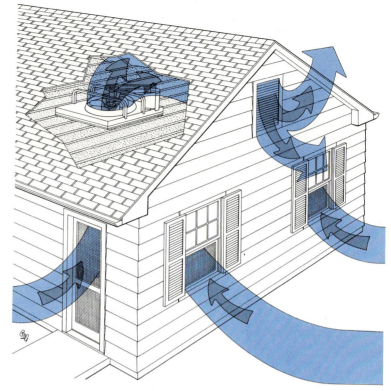

**Sucking in fresh air,** whole-house fan circulates it through lower floors on the way to the attic. Hot, stagnant air is pushed out of the house in the process.

house falling down). Here you use a double shutter, which is designed to work without striking the crossbeam running above it. For all other construction, use the single shutter. You'll have to cut out a ceiling joist, but you'll get better air flow.

## INSTALLATION

The fan should be centrally located at a spot that provides plenty of over-head space (for work and for air flow). Armed with the fan's dimensions, go into the attic and mark the first corner. This must be just inside a joist or truss which will be made part of the basic framing. Using the first mark as a reference, mark off the remaining three corners and drill a pilot hole at each point. Back downstairs, use a

straightedge and join the four drilled holes with a pencil line. The next step is to cut out the ceiling. On a plasterboard ceiling, a linoleum knife will easily cut through.

If you have 2 x 6 or larger wood joists, cut out the middle section and reinforce the sides with 2 x 6 headers. Make the flat-board plate for the fan to rest on out of 1 x 4's (see drawing).

On ceilings with 2 x 4 trusses, do not cut out the middle section. Instead, add two-inch headers at each end and 2 x 6 headers on the sides, notched to fit over the trusses. Make the flatboard plate out of 1 x 6 or 1 x 8 stock.

After nailiing the flat-board plate to the reinforced joists or trusses, lift the fan through the opening into the attic. (A husky friend would be a definite asset at

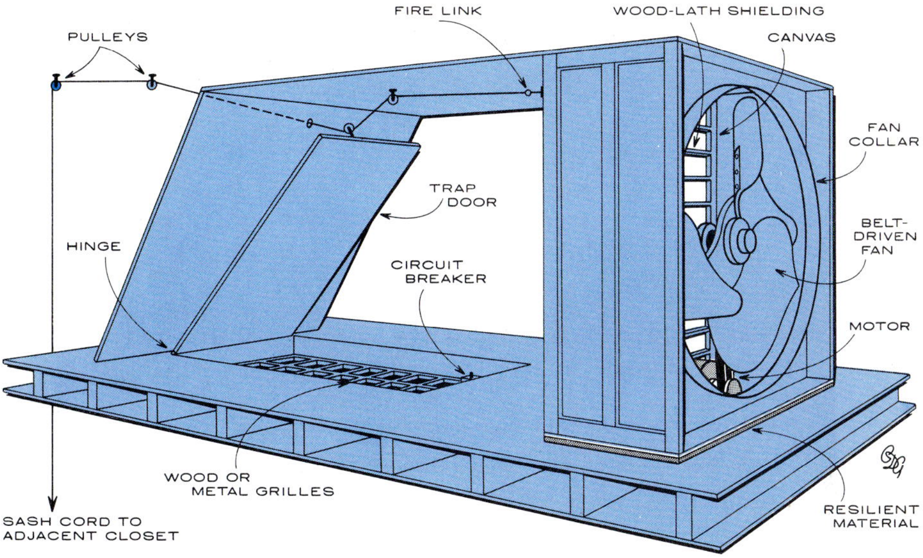
PULLEYS
FIRE LINK
WOOD-LATH SHIELDING
CANVAS
FAN
COLLAR
TRAP
DOOR
HINGE
CIRCUIT
BREAKER
BELT-
DRIVEN
FAN
MOTOR
WOOD OR
METAL GRILLES
SASH CORD TO
ADJACENT CLOSET
RESILIENT
MATERIAL

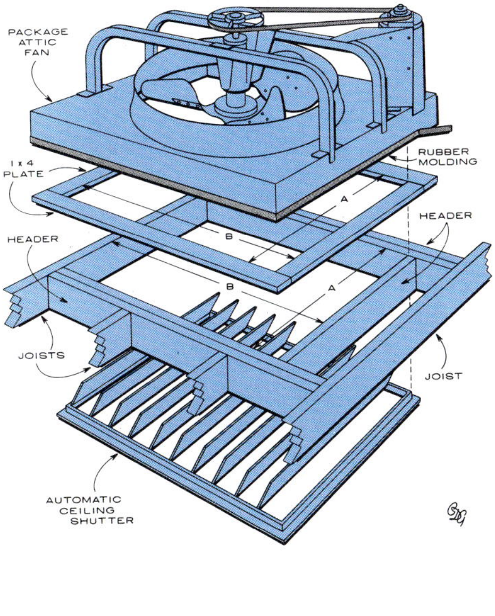
PACKAGE
ATTIC
FAN
RUBBER
MOLDING
1 x 4
PLATE
HEADER
HEADER
A
B
JOISTS
JOIST
AUTOMATIC
CEILING
SHUTTER

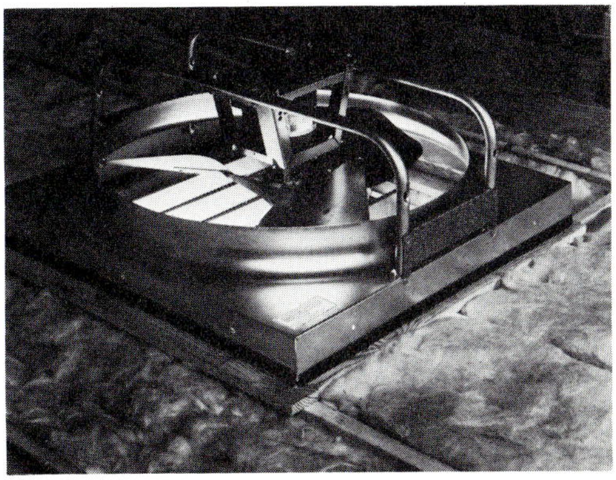

**In vertical mounting,** door in plenum chamber can be lowered over the attic cutout so that the fan (top) acts as an attic fan only. But horizontal mounting (left) is simpler and draws more air.

## Specifications for whole-house fan*

| Fan size | Shutter size overall | Attic exhaust area in square feet | | | | Ceiling framing dimensions | | |
|---|---|---|---|---|---|---|---|---|
| | | Wood louvers | Metal louvers | ½" mesh wire | No. 8 mesh wire | A | B | Plate material |
| 24" | 36⅝" x 32⅛" | 12 | 9 | 7 | 8 | 34⅞" | 30⅜" | 1x4 |
| 30" | 40⅝" x 35½" | 16 | 12 | 10 | 11 | 38⅞" | 33¾" | 1x4 |
| 36" | 43⅞" x 39⅝" | 27 | 21 | 17 | 18 | 42⅞" | 37⅞" | 1x4 |

*Supplied by Hunter Corp.

this point.) Press the rubber molding onto the bottom edges of the fan frame and position the unit squarely over the opening. No anchoring is required. The fan merely rests on the rubber, which seals the edges from air leaks.

Finally, install the shutter, using the specific instructions packed with the type you get, and wire the fan to an AC supply through a switch.

If you've decided to mount your fan vertically in a plenum chamber, all this still applies—and more. Naturally, you'll have to build the chamber, as shown in the drawing. One thing you won't have to do, however, is to take the fan out of its shipping container: Use it as a mounting box.

## OPERATION

Always be sure that at least one window or entry is open in the house before switching on the fan. Today's well-insulated homes seal airtight, forcing the fan to strain and suck air down chimneys. If you own a fireplace, this could mean soot in your living room. (Never operate the fan when there's a fire.) Also, fumes from the hot-water heater or furnace—which should go up the chimney—will be sucked into the house.

# Air controller compensates for negative pressure

## By Evan Powell

Negative pressure. These words describe a costly and potentially dangerous condition in your home. Use any appliance that exhausts air to the outside—anything from a bathroom fan or dryer to your fireplace or furnace—and it draws air from the inside. If your home is "loose," air is sucked in through cracks and windows to make up the difference in pressure. But on well-insulated, or "tight," homes, negative pressure develops: Appliances become starved for air, and that causes inefficient, expensive operation at best and dangerous pollutants at worst.

One cure: Skuttle's Make-Up Air Control. The temperature-operated version mounts on the return-air side of the heating system and connects to an outdoor air supply through four-inch tubing and a screened hood. A bypass tube attaches to the warm-air side of the furnace.

When the furnace blower starts, warm air flows through the bypass tube. And when the temperature rises, an adjustable heat-sensitive coil opens a damper in the intake-air duct. Fresh air flows in and is filtered, heated, and humidified before entering the living area.

According to Greg Costley, engineering manager at Skuttle (Marietta, Ohio 45750), the sizing of the two temperature-controlled units is based on average residential requirements. Model 204, shown on page 360, is for furnaces up to 150,000 Btu; model 206, over 150,000 Btu. The adjustment of the bimetal coil allows further tailoring for individual homes. The lower temperature levels of heat pumps probably won't actuate the unit. For those installations, a barometric control is used (model 216). It can be adjusted to open only if negative pressure exists. All models are available nationally from heating and air-conditioning dealers. No suggested retail prices are available, but in my area I was quoted $50 to $60 for model 216 and $70 for 204 and 206.

**Air-controller kit** contains bypass tubing, fittings, screened intake hood—but not intake tubing. Manual control knob on side of unit can be locked in open position for fresh air when central air conditioner is running. Installation (right) is simple: Use control flange as template and  cut hole with tin snips. Air inside this home seemed noticeably fresher after unit was installed.

# BIG
# 10
# CHECKLIST

Here's where it all pays off. In the preceding chapters, you have seen how to initiate the modifications and use today's technological advances in home heating equipment to reduce your home's operating costs and make it more comfortable as well. But just in case you overlooked something, look through the checklist. You might stumble upon a good weekend project there.

And don't let it stop here. In all your future home planning, keep an eye peeled for the best way and best equipment to do the job efficiently. The recent innovations are exciting, especially after years of seemingly slight improvement. Still, they're just the beginning of even better things in the realm of home energy equipment.

361

# Big 10 checklist for home energy saving

## By Paul Bolon

Twenty industrialized countries are each trying to cut oil imports and, thereby, reduce inflation. The target of efforts in this country is residential consumption, which accounts for 22 percent of U.S. energy use.

Campaigns to reduce oil imports are not new, of course. But not enough of us have implemented basic conservation repairs and improvements—steps that will also put money in homeowners' pockets.

As you will see when you read the checklist, you will find how to implement most of these improvements—and all of the major ones—in this book.

The Department of Energy has developed the following checklist to encourage Americans to complete every simple energy-saving step. The goal is to reach all 74 million residences in this country. The checklist's first six tips are low-cost corrections. Improvements entitling homeowners to tax credits are printed in **boldface**.

## 1 WEATHER-STRIPPING/ CAULKING

Check for cracks around doors, windows, and other openings. Seal cracks with **caulking** or **foam sealant**, apply **weather-stripping** around doors and window sashes. Air escaping through cracks results in heat loss.

## 2 THERMOSTAT SETBACK

Set thermostats at 65 degrees in winter and at least five degrees lower at night or when away. (Higher settings are recommended for the elderly or infants, or when sick.) In summer, temperature should be no lower than 78 degrees. Consider buying a **computerized thermostat**.

### 3
### WATER
### HEATERS

Water heaters are major energy consumers. Lower thermostat setting (to 120 degrees) and consider adding an **insulation wrap kit**. Install water flow restrictors in shower and taps; they cut hot-water use without affecting comfort. If replacing a water heater, choose a heat pump, an energy-efficient model or a **solar unit.**

### 4
### HEATING/
### COOLING
### SYSTEM

Clean or replace dirty filters. Close vents in unused rooms and add **insulation** to ducts and pipes in unheated spaces. Consider devices that can increase the efficiency of your existing system: more efficient **replacement burners** or **automatic vent dampers**.

### 5
### SUNLIGHT

In summer, shade sunny windows with shades, awnings, drapes, or **solar films or screens**. Unshaded south windows yield a net heat gain in winter.

### 6
### APPLIANCES/
### LIGHTING

Fully load dryer and clothes- and dishwashers. When replacing them, buy energy-efficient models. Turn off unnecessary lights; use fluorescent lamps where possible.

### 7
### ATTIC
### INSULATION

Compare present attic insulation with recommended levels; add **insulation** if necessary. Be sure attic door is well-sealed and insulated, too.

### 8
### FLOORS AND
### FOUNDATION
### WALLS

Foundation walls or floors over crawl spaces and unheated basements should be insulated. Check between floor joists and install **insulation** to suggested levels.

### 9
### WINDOWS
### AND DOORS

**Storm windows and doors, double- or tri-ple-pane windows,** and insulating shades will reduce loss of heat or conditioned air at glazed areas.

### 10
### EXTERIOR
### WALLS

When remodeling or re-siding your home, consider adding extra **insulation** inside or under siding.

## TAX CREDITS

Many modifications to improve your home's heating and cooling characteristics qualify for federal and state tax credits. The amounts and terms of these credits vary each year, so you should check with your tax office for details; they may be sufficient, however, to influence your decision about implementing the improvements.

# Index

## A

Acoustical leak detection, 183–185
ACRI, *see* Air Conditioning and Refrigeration
    Institute
Active solar systems, 255
Advanced Mechanical Technology, Inc., 324
Air circulation for wood stove heat, 129–132
Air conditioners
    computer-controlled, 22–23
    waste-heat water heater, 314–317
    *see also* Heat pumps
Air Conditioning and Refrigeration Institute
    (ACRI), 63
Air Control Systems, 280
Air Floor, 280
Air leaks, 183–185, 186–190
Air pollution
    from catalytic wood stoves, 13
    heat exchangers for, 343–349
    in tight house, 340–342
Air-to-air heat exchanger, 343–349
Allaire, Roger, 13
Allied Chemical, 42
Amana Heat Transfer Module, 37
American Society for Testing and Materials
    (ASTM), 157
Amtrol, Inc., 324
Appalachian Regional Commission, 248
Applied Ceramics catalytic wood stove, 13
Arthur D. Little, Inc., 35, 282, 316–317
ASTM, *see* American Society for Testing and
    Materials
Auburn University, Wood Heating
    Laboratory, 15
Automatic flue dampers, 48–49
Autotronics thermostat, 81

## B

Babington, Robert, 32–33
Barnes, Randy, 277–279
Barnett, Allen, 270
The Barrier kit storm window, 233
Basements, 215–217
Batelle Institute, 62
Batey, John, 26–28, 30, 33
Benson, Doug, 110
Besant, R. W., 344
Better Heating-Cooling Council, 65
Bio-Energy Systems, Inc., 290
Birmingham Ponderosa Catalytic stove, 14
Bissett, Donald, 140
Blue-flame technology, 30–31
Blueray oil burner, 31
Boardman, Brian, 276
Boardman Energy Systems, 275–276
Boiler/radiator fireplace system, 105–106
Boilers
    blue-flame technology, 30–31
    low-volume gas, 36–37
    pulse combustion, 3, 33, 38–40
Bond, Ted, 246
Braun, Tom, 71
Brookes, John, 192
Brookhaven National Laboratory, 26–30
Brownell, Jim, 65
Brunberg, Ernst-Ake, 73
Buchanan, David, 210
Builders Collaborative, Inc., 293

## C

Calcium chloride hexahydrate, 269
Callor, John, 57